迟滞非线性系统
辨识与智能控制

高学辉◎著

吉林大学出版社

·长春·

图书在版编目（CIP）数据

迟滞非线性系统辨识与智能控制/高学辉著．——长春：吉林大学出版社，2022.7
ISBN 978-7-5768-0017-3

Ⅰ．①迟… Ⅱ．①高… Ⅲ．①迟滞—非线性控制系统—研究 Ⅳ．①TP273

中国版本图书馆 CIP 数据核字（2022）第 132764 号

书　　名：迟滞非线性系统辨识与智能控制
　　　　　CHIZHI FEIXIANXING XITONG BIANSHI YU ZHINENG KONGZHI

作　　者：高学辉　著
策划编辑：李伟华
责任编辑：甄志忠
责任校对：张文涛
装帧设计：中北传媒
出版发行：吉林大学出版社
社　　址：长春市人民大街 4059 号
邮政编码：130021
发行电话：0431-89580028/29/21
网　　址：http://www.jlup.com.cn
电子邮箱：jldxcbs@sina.com
印　　刷：廊坊市海涛印刷有限公司
开　　本：710mm×1000mm　　1/16
印　　张：9
字　　数：110 千字
版　　次：2023 年 7 月　第 1 版
印　　次：2023 年 7 月　第 1 次
书　　号：ISBN 978-7-5768-0017-3
定　　价：45.00 元

版权所有　翻印必究

前　言

迟滞是常见的非线性现象，如电磁系统、智能材料、继电器等都具有典型的迟滞特性，在某些情况下可以利用迟滞现象为我们服务。比如，利用电压比较器的迟滞特性产生周期性振荡电路从而作为信号发生器使用等。但大多数情况下，迟滞的存在会影响系统性能、降低系统精度，需要克服迟滞特性的影响。

本书针对 Preisach 迟滞模型、Backlash-like 迟滞模型、Prandtl-Ishlinskii 迟滞模型和 Bouc-Wen 迟滞模型描述的非线性系统，研究了系统辨识、状态观测和智能控制器设计等问题，为迟滞非线性系统辨识与智能控制提供了新思路和新方法。

第1章绪论，介绍了本研究的背景及意义，迟滞模型的分类及各类迟滞模型的发展历史及研究现状，Hammerstein 系统模型的定义及研究现状，以及迟滞 Hammerstein 系统目前存在的问题和研究方向，并给出了本书的研究内容。

第2章考虑 Preisach 迟滞 Hammerstein 系统在仅仅已知输入输出情况下，提出了基于盲辨识的 Hammerstein 系统辨识方法，该方法仅需要输入输出数据已知，而不需要已知系统阶次，放松了辨识条件，使其更具一般性。针对其 Preisach 迟滞模型辨识方法需要进一步研究的问题，提出了一种新的下三角矩阵的辨识方法，可在线辨识 Preisach 模型并避免了模型"擦除"特性的影响；在此基础上，采用滑模控制和逆模型控制结合的混合控制策略，提高了控制系统收敛速度和鲁棒性。

第 3 章仍然对迟滞 Hammerstein 系统进行研究，讨论了基于 Backlash 类模型的迟滞 Hammerstein 系统的规定性能控制问题，简化了现有规定性能函数，使其更加简便实用；提出了在控制器设计过程中将矢量误差转化为标量误差，再转化为规定性能误差，以简化控制器设计。但是误差转换会带来精度降低的问题，因此采用简化的规定性能函数保证闭环系统的控制精度。然后提出了模型参考自适应控制策略对 Backlash 类模型描述的迟滞 Hammerstein 系统进行控制；应用 Lyapunov 函数保证控制系统稳定且引进 Lambert-W 函数证明所提出的控制策略稳定且暂态和稳态误差均收敛于规定性能范围内。

第 4 章研究 Prandtl-Ishlinskii 模型描述的迟滞非线性系统的辨识问题，提出了采用 Hopfield 神经网络辨识迟滞非线性系统。采用正交阵等变换将迟滞非线性系统变换为正交标准型，设计 Hopfield 神经网络辨识系统参数并验证了所提出方法的有效性。

第 5 章对 Bouc-Wen 模型描述的迟滞非线性永磁同步电机系统进行研究，首先在考虑磁滞损失和其他非线性影响下，建立了永磁同步电机系统的非线性状态空间模型；其次，为了简化控制器设计，把所建立的模型转化为标准形式。由于模型部分状态未知，所以提出了基于回声神经网络的扩张状态观测器估计未知系统状态。根据所得系统状态估计，提出了神经网络滑模控制器控制迟滞非线性系统；最后，为了降低滑模控制抖震的影响，采用双曲函数代替符号函数并保证了系统的稳定性。

  本书部分内容总结了笔者博士期间的研究成果，另外部分内容为近年来研究的新进展，可作为控制科学与工程学等专业硕士、博士及相关工程技术人员参的考书。

  由于本人水平有限，书中难免会出现错误或者不妥之处，敬请读者指正！

高学辉

2021 年 9 月

目 录

第 1 章 绪 论 1
  1.1 迟滞非线性系统概述 1
  1.2 迟滞非线性辨识和自适应控制研究现状 5
  1.3 迟滞非线性系统辨识与控制研究中存在的问题 11
  参考文献 13

第 2 章 Preisach 迟滞非线性系统下三角矩阵分段一致辨识与滑模控制 31
  2.1 问题的提出 31
  2.2 研究内容 32
  2.3 问题描述 34
  2.4 Hammerstein 系统辨识 35
  2.5 混合控制器设计 44
  2.6 仿真与实验 49
  2.7 结论 55
  参考文献 55

第3章 Backlash-like 迟滞非线性系统预设精度自适应控制 59
3.1 引言 59
3.2 问题描述 61
3.3 控制器设计 63
3.4 仿真 71
3.5 结论 77
参考文献 77

第4章 Prandtl-Ishlinskii 迟滞非线性系统的 Hopfield 神经网络辨识与自适应控制 81
4.1 引言 81
4.2 问题描述 83
4.3 Hopfield 神经网络辨识 85
4.4 自适应控制器设计 89
4.5 仿真 90
4.6 结论 94
参考文献 95

第5章 Bouc-Wen 迟滞非线性系统扩张状态观测器与回声神经网络滑模控制 99
5.1 引言 99
5.2 问题表述 103
5.3 回声神经网络 108
5.4 扩张状态观测器 109
5.5 滑模控制器设计 113

5.6 仿真 118
5.7 总结 126
参考文献 127
致 谢 133

第1章 绪 论

1.1 迟滞非线性系统概述

迟滞也被称为滞回,与单纯时间意义的滞后不同,迟滞不仅在时间上表现为滞后,在空间上同样存在滞后现象,是一种典型的非线性环节。迟滞是指一系统的状态不仅与当前系统输入有关,还会随着过去输入路径的不同而不同。因此,迟滞系统即使输入相同,在不同路径上也会得到不同的系统输出,典型迟滞现象如图1.1所示。常见的迟滞现象有磁滞现象[8-14]、弹性迟滞现象[18-20]等。

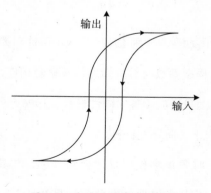

图 1.1 典型迟滞非线性环节

"迟滞（Hysteresis）"一词源于古希腊语"υστ ερησις"，于1890年由西尔·詹姆斯·阿尔弗雷德·尤因描述磁性材料时在希腊语基础上创造得到。随着对迟滞研究的深入，各种迟滞模型被提出；从建模方法来看，目前迟滞模型可以粗略地分为两大类：物理模型和唯象模型。常见的物理模型有 Jiles-Atherton 模型[21-24]，而唯象模型包括 Preisach 模型[25-32]、Prandtl-Ishlinskii 模型（简称 P-I 模型）[33-40]、Bouc-Wen 模型[41-45]等，如图1.2所示。

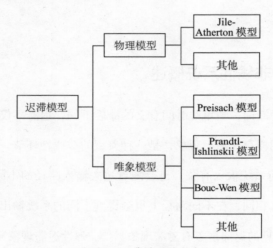

图1.2 迟滞模型的分类

Jiles-Atherton 模型由西莱斯·阿瑟顿于1984年针对磁滞材料首先提出[46]。该模型属于物理模型，模型参数与材料物理特性紧密相连，并且可以计算最大和最小迟滞环。最初该模型只适应于各向同性材料，1996年拉梅什等学者扩展了该模型[47]，使其同时适应于各向异性材料。

与物理模型不同，唯象模型仅仅描述迟滞现象，与具体物理参数无关，因此，唯象模型应用更加广泛。经典的唯象模型是 Preisach 模型[48]。该模型最初为物理模型而且也不是由普莱扎赫提出的[49]，而是由韦斯和德弗洛伊德里希[50]于1916年研究铁磁体模型时首先提出的，1935年费伦斯·普罗萨奇[48]对该模型做出几何解释，20世纪70年代，苏联数学家马克·克拉斯诺塞尔斯基和博布罗夫斯基[51]发现Preisach模型包含通用的数学思想，从而将Preisach模型从物理模型中分离出来，通过Preisach算子将其改进为纯粹的唯象模型，并命名为Preisach模型。

另一种应用广泛的唯象模型是P-I模型，1928年普兰特[52]提出了流体力学的弹性可塑性模型，1954年，伊什林斯基[53]在此基础上发展为迟滞模型。20世纪70年代，马克·克拉斯诺塞尔斯基和博布罗夫斯基[51]将该模型命名为Ishlinskii模型。1994年，奥古斯托·维辛丁[49]将由Play和Stop算子构成的一大类该模型统一命名为Prandtl-Ishlinskii模型。

Bouc-Wen模型最早是一类应用广泛的描述迟滞的物理模型，该模型于1971年由布克提出[54]，1976年温义桂[55]扩展了该模型，使其成为准唯象模型，适用性更加广泛。与Jiles-Atherton模型不同，近年来，随着研究的深入，Bouc-Wen模型参数逐渐摆脱其物理意义，特别是费卡尔·伊霍瓦内等学者[43]于2007年证明了Bouc-Wen迟滞系统的有界输入有界输出（BIBO）稳定性以后，Bouc-Wen模型彻底演进为唯象模型。因此可适用于磁滞阻尼器、土木建筑阻尼迟滞、钢结构迟滞、多向铰接迟滞等多方面。

对唯象模型来说，由于与物理参数无关，仅仅描述了迟滞现象，因此应用范围广泛，通用性好，只要具有迟滞特性均可应用唯象模型描述，所以，唯象迟滞模型一直是研究的重点。

从模型动态特性来看，迟滞模型还可以分为频率相关（Rate-Dependent）和频率无关（Rate-Independent）两类。早期迟滞模型大多为频率无关模型，如 Preisach 模型、P-I 模型等。随着研究的深入，迟滞模型多数可描述频率相关迟滞现象，如近几年改进的 Preisach 模型等。所谓频率相关是指迟滞输入减弱为零后，迟滞输出相应仅持续有限时间，即其记忆性为有限的，随着输出降为零而消失；换句话说，频率相关迟滞相位滞后与输入频率有关，随着频率的降低而趋向于零。而频率无关迟滞现象在瞬态消失后仍具有长久的记忆性，迟滞输出响应只与过去输入路径有关，而与过去输入路径的速度（频率）无关。目前来看，随着技术的进步和模型精度要求的提高，频率相关迟滞已经引起了广泛的重视，也是迟滞研究的热点之一。

迟滞广泛存在于电机、智能材料、电磁执行器等系统中，如伺服电机[58-61]、形状记忆合金[62-65]、磁滞执行器[66-70]等均具有典型的迟滞特性。除此之外，生物医学系统、社会系统和声波等系统中也具有典型的迟滞特性。

随着科技的进步与社会的发展，对系统的研究越来越精确，指标要求也越来越高。以前可线性化处理的迟滞特性随着精度要求的提高在许多系统中变得越来越不可忽略，如生物医学模型中迟滞的影响、社会管理模型中政令的明显的迟滞效应等在新的精确模型下不可简单线性化。在工程应用中，该现象更为明显。比如高精度伺服电机系统、机器人系统、高精度压电陶瓷驱动器等均需要精确建立迟滞模型以提高控制精度。特别是微纳级的驱动器、执行器更是国内外研究的重点，因此，对迟滞的精确建模、辨识和控制问题是该类系统研究的核心问题。

Hammerstein 系统是一类典型的非线性系统，由非线性环节串联线性环节构成。系统输入 u 是 Hammerstein 系统非线性环节输入,既是非线性环节输出同时又是线性环节输入,线性环节输出即系统输出为 y,系统模型可描述为:

$$\begin{cases} \dot{x}(t) = Ax(t) + B\upsilon(t) \\ y(t) = Cx(t) \\ \upsilon(t) = f(u(t)) \end{cases} \quad (1.1)$$

其中,$\upsilon(t) = f(u(t))$ 表示非线性环节,该环节非线性结构可以已知,也可以未知。Hammerstein 系统属于模块化结构系统的一种,模块化结构系统包括 Hammerstein 系统[32,71]、Wiener 系统[72,73]和 Sandwich 系统[74-76]。由于模块化结构系统可以十分方便地描述死区、迟滞、摩擦、饱和等非线性环节,同时也可以十分方便地表达由智能算法描述的结构未知的非线性环节,因此一直广受国内外研究学者的关注,在非线性研究和工程实践中有十分重要的作用。

1.2 迟滞非线性辨识和自适应控制研究现状

迟滞目前广泛应用于工程、材料、生物、经济等领域,例如控制系统,电路、智能材料、机电系统、磁滞材料、免疫学、细胞与生物遗传学、神经学、博弈和资本管理、社会管理等方面都有大量迟滞研究应用论文。一般来说,迟滞物理模型多用于材料和生物方面,唯象模型多用于磁滞、智能材料、机电系统、经济管理等方面,特别是工程研究中唯象模型应用越来越广泛。相对而言,唯象模型由于描述的是迟滞的现象,通过数学表达,因此更加具有广泛性,是研究的重点。

迟滞的研究从 20 世纪 60 年代开始预热,80 年代以后研究逐渐增多。随着科学技术的进展,特别是计算机技术的飞速发展,使得人们对模型精度的要求越来越高，研究工具也越来越先进，因此，进入21世纪以来，迟滞研究迎来了发展的时期，研究人员越来越多。即便如此，由于迟滞的复杂性和多值性，仍有许多问题亟待解决，如模型辨识的问题，精确控制问题等。

近年来，不论是对物理模型还是唯象模型研究，均取得了一定进展，但从长远来看，唯象模型因其广泛的应用性将成为迟滞研究的重点方向，下面对唯象迟滞模型的辨识与控制研究现状简要论述。

1.2.1　Preisach迟滞模型研究现状

自20世纪70年代苏联数学家马克·克拉斯诺塞尔斯基和博布罗夫斯基[51]将Preisach模型改进为唯象模型后，Preisach模型一直深受关注。90年代，梅尔戈兹[1]提出了一阶转换曲线分段输出辨识法辨识Preisach模型，使得Preisach模型的应用前进了一大步。2005年，谭小波和巴拉斯[77]针对Preisach模型提出了两种类最小二乘的递归辨识算法，一种基于迟滞输出，另一种基于迟滞输出的时间差分，并证明了该算法的持续激励（PE）条件。在此基础上，根据辨识结果提出了一种自适应逆模型控制并通过仿真和实验验证。法拉和莫哈尼[78]针对各向同性矢量Preisach模型，应用新的导出方程，通过辨识相应的标量模型导出矢量模型并验证了该方法的正确性。鲁德曼和伯特伦[79]采用最小二乘法辨识离散动态Preisach模型的Preisach密度函数，并在有限迟滞数据下求其全局解。笔者[80]在前人研究的基础上针对Preisach模型提出了下三角矩阵分段一致辨识方法，取得了不错的结果。

Preisach模型控制可以粗略地分为逆模型补偿控制和无逆模型控制两类。宋钢兵等学者[81]对微定位压电陶瓷执行器采用逆 Preisach 模型补偿控制，并采用 PD 反馈控制处理压电陶瓷执行器的动态误差。与此类似，包康圭和金群安[82]针对形状记忆合金执行器设计了前馈与反馈结合的控制策略，前馈环节由模糊逆 Preisach 模型构成来补偿迟滞非线性，反馈环节由 PID 调节系统输出与期望输出的误差，实际实验结果验证了该方案的有效性。同样，刘磊等学者[83]设计了类似的复合控制策略控制压电陶瓷执行器。其前馈为逆 Preisach 模型（且 Prieach 模型首先由最小二乘法辨识得到）补偿迟滞非线性，反馈由二阶 PI 控制器抑制扰动的鲁棒性。实验可以验证该控制策略精确有效。

经典的 Preisach 模型为频率无关模型，而大多数智能材料、磁滞执行器等频率相关，肖顺利和李杨民[85]针对频率相关压电陶瓷执行器，在经典 Preisach 模型基础上，提出了改进的频率相关 Preisach 模型，首先应用最小二乘法辨识经典 Preisach 密度函数，然后应用快速傅立叶变换选择合适 Preisach 密度函数补偿实际压电陶瓷迟滞非线性并通过了实验验证。

1.2.2　Prandtl-Ishlinskii 迟滞模型研究现状

直到 21 世纪初学者们才逐渐对 Prandtl-Ishlinskii 模型（简称 P-I 模型）重新感兴趣并对其进行了广泛研究，与 Preisach 模型类似，P-I 模型的控制策略也可以分为两类：逆 P-I 模型控制和无逆 P-I 模型自适应控制。洪维德等学者[87]提出了一种改进的频率相关 P-I 模型处理不同驱动频率下压电陶瓷执行器的控制问题，采用逆模型补偿压电陶瓷迟滞非线性并通过实验验证了该方法的有效性。但是 P-I 模型在其迟滞曲线梯度非正定时，其逆模型不存在，即当曲线斜率为负时存在病态条件，谭小波等学者[88]针对此病态问题，提出了扩展的 P-I 改进模型并采用与文献[87]类似的逆模型控制完成对迟滞非线性的补偿控制。陈新凯等学者[89]提出了一种隐式逆 P-I 模型补偿迟滞非线性，并在此基础上采用自适应控制策略控制连续的迟滞非线性系统。而刘思宁等学者[90,91]采用与文献[89]类似逆 P-I 模型迟滞自适应控制并严格证明了控制策略的稳定性。李志福等学者[92]仍然采用逆 P-I 模型鲁棒自适应控制补偿迟滞非线性，解决了离散 P-I 模型问题，并用双曲正切函数代替符号函数以降低抖震且证明了该闭环系统所有信号最终一致有界（UUB）。谷国迎等学者[93]针对压电陶瓷执行器提出了一种改进的非对称 P-I 模型，与传统改进 P-I 模型不同，该模型并未用复杂的非线性算子代替经典 Play 或 Stop 算子，而是采用广义输入函数代替经典 P-I 模型的线性输入，因此可用较少的参数描述改进的 P-I 模型；在此基础上，采用逆 P-I 模型补偿迟滞非线性并对压电陶瓷执行器进行精确控制。阿尔贾奈德和克雷奇[94]采用纯前馈开环频率相关离散逆 P-I 模型补偿压电陶瓷微定位执行器的迟滞非线性，该模型适用的压电陶瓷微定位执行器频率激励范围为 0.05～100Hz。与前述文献均不同，秦岩丁等学者[95]并未直接计算 P-I 模型逆模型，而是采用高阶多项式代替频率相关 P-I 模型逆模型，其高阶多项式参数可通过实验测得，该方法可以避免求频率相关逆 P-I 模型时的复杂计算，并且取得了较好的控制精度。

无逆 P-I 模型的迟滞自适应控制策略也被深入研究，苏春翌等学者[96,97]对一类用 P-I 模型描述的迟滞非线性系统，设计了一种无需逆 P-I 迟滞模型的自适应变结构控制，并证明了该控制策略的全局稳定性。针对一类纯反馈非线性动态未知的迟滞系统，张秀宇和林岩[98]提出了自适应神经网络动态面控制，采用饱和 P-I 模型描述迟滞非线性，并用性能函数约束跟踪误差，放松了动态不确定性的假设条件，避免了反步法（back-stepping）设计的维数灾难问题，并使得控制率设计简化，降低了计算负担。任贝贝等学者[99,100]对一类用 P-I 模型描述的迟滞输入的纯反馈非线性系统，设计了自适应神经网络反步法控制策略处理该非光滑迟滞的非仿射问题，并采用李雅普诺夫方法证明了该控制系统的一致稳定性且跟踪误差收敛于一个小邻域内。

除逆 P-I 模型控制与无逆 P-I 模型自适应控制策略外，近年来，有逆 P-I 模型的自适应或者其他变结构、智能控制策略也得到了广泛关注。理查德等学者[102]研究了一类形状记忆合金的迟滞非线性控制策略，该控制策略不但采用逆 P-I 模型补偿迟滞非线性，而且采用在线自适应参数调节以提高控制精度并通过实验验证了该控制策略的有效性。阿斯切曼和辛德尔[103]针对气动肌肉的力量特性，采用三种不同模型进行比较，分别为 Bouc-Wen 模型、准静态 Maxwell 滑动模型和 P-I 模型，使用自适应反步法控制带逆模型的气动肌肉系统并取得了较好的控制效果。而巴沙什和亚利利纳德[104]则采用滑模与逆 P-I 模型相结合的控制策略控制压电陶瓷微纳执行器迟滞系统，针对带迟滞非线性的二阶动态系统，设计逆 P-I 模型补偿迟滞非线性，且利用带摄动估计的滑模控制器控制该系统以提高控制精度和动态特性。

1.2.3 Bouc-Wen 迟滞模型研究现状

Bouc-Wen 模型真正演进为唯象模型是近年来才发生的，之前大多作为物理模型或准唯象模型研究。费卡尔·伊霍瓦内等学者[43]于 2007 年证明了 Bouc-Wen 模型的迟滞系统有界输入有界输出（BIBO）稳定后，Bouc-Wen 模型演进为唯象模型。穆罕默德·伊斯梅尔等学者[105]首次明确说明了 Bouc-Wen 模型是物理模型和唯象模型的统一，在本书中，基于综合考虑后，将 Bouc-Wen 模型彻底归类为唯象模型。但是由于 Bouc-Wen 模型参数较多，辨识复杂，较难应用，所以一开始并没有引起学者们的足够重视。近年来，随着对 Bouc-Wen 模型辨识研究的深入，Bouc-Wen 模型逐渐成为一类重要的描述迟滞非线性的模型。

目前，学者们针对 Bouc-Wen 模型的辨识和控制问题进行了新的研究，劳达尼等学者[4]提出了一种新的混合启发式辨识算法标准拓扑进化最优算法辨识 Bouc-Wen 模型，该算法由三种不同粒子群优化算法协作完成辨识 Bouc-Wen 模型。徐延海等学者[106]针对汽车侧面稳定通过磁流变阻尼器对可变刚度和阻尼的悬挂系统进行研究，用 Bouc-Wen 模型对磁流变阻尼器进行建模并采用模糊控制器控制该系统。而哈比内扎等学者[107]则扩展了经典 Bouc-Wen 模型将其应用于多自由度迟滞模型，并给出了辨识方法和逆模型补偿控制方法。刘智等学者[108]则对用 Bouc-Wen 描述的迟滞非线性系统采用自适应神经网络输出反馈控制策略控制系统，并且系统状态无法直接测量，设计了鲁棒滤波器处理系统状态约束，并证明了控制系统的半全局最终一致稳定。

1.2.4　Backlash 迟滞模型研究现状

Backlsah 类迟滞模型（Backlash-Like Hysteresis）也是被广泛研究的一类唯象迟滞模型，苏春翌等学者[110]针对一类未知的 Backlash 类迟滞动态系统提出了鲁棒自适应控制，采用微分方程描述 Backlash 类迟滞非线性，并应用鲁棒自适应控制算法保证该自适应系统的全局稳定性和跟踪误差收敛于期望精确度内。并在此基础上进一步提出了结合模糊通用函数近似的自适应控制器[111]，并设计了无逆迟滞模型的自适应控制。周晶等学者[112]在苏春翌等学者[106,107]研究基础上，讨论了干扰下的 Backlash 类迟滞非线性系统的控制问题，并采用反步法设计自适应控制器并取得了很好的控制效果。而王焕钦等学者[113]研究了随机迟滞非线性非严格反馈系统的控制问题，为了处理非严格反馈的变量，引入了变量分离技术，并设计了神经网络自适应反步法控制器，而且证明了该控制策略的半全局一致有界。黄晓宇等学者[114]针对汽车线控转向、驱动和刹车问题进行研究，并将操控系统中的迟滞作为干扰采用 Backlash 类迟滞模型描述，设计了加权增益调度状态反馈控制并通过了实验验证。刘烨和林岩[115]则考虑了系统参数和状态均未知时 Backlash 类迟滞非线性系统的控制问题，采用反步法和高增益观测器观测系统状态并设计控制器并使其达到预设的 L_∞ 跟踪性能。

1.3　迟滞非线性系统辨识与控制研究中存在的问题

从迟滞非线性系统的研究现状可以看到，经过共同努力，迟滞非线性系统辨识和智能控制取得了一系列重要成果。但是仍然有一些关键而重要的问题亟需突破。主要包括：

1.3.1 Preisach 迟滞模型新的辨识方法

Preisach 迟滞模型成熟的辨识方法有多种，但是这些方法要么难以求得精确解，要么受"擦除"特性的影响，寻求一种既不受"擦除"特性影响又容易计算的新辨识方法是十分必要的。

1.3.2 系统参数所知甚少的情况下迟滞系统辨识

迟滞系统辨识研究由来已久，并取得了大量的成果。但是这些大多是在各种假设条件下，或者需要知道大量系统特征的前提下进行的，如何在少量系统特征已知情况下辨识迟滞非线性系统仍然是亟需解决的问题。

1.3.3 系统状态未知的迟滞非线性系统的控制

如前所述，针对系统状态未知的迟滞非线性系统控制问题虽然取得过部分成果，但是总体来看还需要继续深入研究。特别是系统状态未知、参数也未知情况下的迟滞非线性系统的控制问题，需要给予更多关注。

1.3.4 迟滞非线性系统的规定性能控制

目前绝大多数迟滞非线性系统控制研究集中于各种控制策略研究，在设计控制器过程中为了简化控制器设计往往会造成误差的扩散。这就形成了一对矛盾：追求控制精度势必会让控制器越来越复杂，但是简化控制器又会降低控制精度。本书作者先前针对此矛盾提出了预设精度控制既保证了控制精度，又在一定程度上简化了控制器设计，但是此问题仍需进一步深入研究。

参考文献

[1] MAYERGOYZ ID. Mathematical models of hysteresis [M].New York: Springer-verlag, 1991.

[2] MATSUO T. SHIMASAKI M. Identification of a Generalized 3D Vector Hysteresis Model Through the Superposition of Stop- and Play-Based Scalar Models [J]. IEEE Transactions on Magnetics, 2007,43(6):2965-2967.

[3] QIN R, RAHMAN MA. Magnetic equivalent circuit of PM hysteresis synchronous motor [J]. IEEE Transactions on Magnetics, 2003, 39 (5): 2998-3000.

[4] AUDANI A, FULGINE FR, SALVINI A. Bouc–Wen Hysteresis Model Identification by the Metric-Topological Evolutionary Optimization [J]. IEEE Transactions on Magnetics, 2014,50(2):621-624.

[5] GARSHELIS IJ, GREVECOEUR G. Nondestructive Detection of Inhomogeneity in the Magnetic Properties of Materials With a Moving Mag-net Hysteresis Comparator [J]. IEEE Transactions on Magnetics, 2012,48(11):4409-4412.

[6] HONG KK, SUN KH, HYUN KJ. Analysis of hysteresis motor using finite element method and magnetization dependent model [J]. IEEE Transactions on Magnetics, 2000, 36(4): 685-688.

[7] HAUSER H, MELIKHOV Y, JILES DC. Examination of the Equivalence of Ferromagnetic Hysteresis Models Describing the Dependence of Magnetization on Magnetic Field and Stress [J].IEEE Transactions on Magnetics, 2009,45(4):1940-1949.

[8] CINCOTTI S, DANERI I. A PWL circuit approach to the definition of an approximation model of scalar static hysteresis [J] . IEEE Transactions on Circuits and Systems I: Fundamental Theory and Applications, 2002,49(9):1290-1308.

[9] ROBRET PA, ROTKIN SV. Modeling hysteresis phenomena in nanotube field-effect transistors [J] . IEEE Transactions on Nanotechnology, 2005,4(2):284-288.

[10] MAO HF, YANG X, CHEN ZL, et al. A Hysteresis Current Controller for Single-Phase Three-Level Voltage Source Inverters [J] . IEEE Transactions on Power Electronics, 2012,27(7):3330-3339.

[11] CAO L, LI G. Complete Parallelogram Hysteresis Model for Electric Machines [J] . IEEE Transactions on Energy Conversion, 2010,25(3):626-632.

[12] HSU JT, NGO KDT. Application of field-based circuits to the modeling of magnetic components with hysteresis [J] . IEEE Transactions on Power Electronics,1997,12(3):422-428.

[13] SUUL JA, LJOKELSOY K, MIDTSUND T, et al. Synchronous Reference Frame Hysteresis Current Control for Grid Converter Applications [J] . IEEE Transactions on Industry Applications, 2011,47(5):2183-2194.

[14] HOLMES DG, DAVOODNEZHAD R, MCGRATH BP. An Improved Three-Phase Variable-Band Hysteresis Current Regulator [J] . IEEE Transactions on Power Electronics, 2013,28(1):441-450.

[15] ZHANG HW, AHMAD S, LIU GJ. Modeling of Torsional Compliance and Hysteresis Behaviors in Harmonic Drives [J]. IEEE/ASME Transactions on Mechatronics, 2015, 20(1):178-185.

[16] KAR S, SARANGI SK, RAO VV. A Comparative Study on Hysteresis Losses in High Tc Tapes for Superconducting Fault Current Limiter Applications [J]. IEEE Transactions on Applied Superconductivity, 2013,23(3):8200404.

[17] LEANG KK, DEVASIA S. Feedback-Linearized Inverse Feed-forward for Creep, Hysteresis, and Vibration Compensation in AFM Piezo-actuators [J]. IEEE Transactions on Control Systems Technology, 2007,15(5):927-935.

[18] DUTTA SM, GHORBEL FH. Differential hysteresis modeling of a shape memory alloy wire actuator [J]. IEEE/ASME Transactions on Mechatronics, 2005,10(2):189-197.

[19] RICCARDI L, NASO D, TURCHIANO B, et al. Adaptive Control of Positioning Systems with Hysteresis Based on Magnetic Shape Memory Alloys [J]. IEEE Transactions on Control Systems Technology, 2013,21(6):2011-2023.

[20] ABDO B, ARBELSEGEV E, SHTEMPLUCK O, et al. Observation of Bifurcations and Hysteresis in Nonlinear NbN Superconducting Microwave Resonators [J]. IEEE Transactions on Applied Superconductivity, 2006,16(4):1976-1987.

[21] XU MX, CHEN ZH, XU MQ, et al. Discussion of modified Jiles-Atherton model including dislocations and plastic strain [J]. International Journal of Applied

Electromagnetics and Mechanics, 2015,47(1):61-73.

[22] ROBERT M, JACEK I. The frequency-dependent Jiles–Atherton hysteresis model [J]. Physica B: Condensed Matter, 2015(463):68-75.

[23] HAMIMID M, FELIACHIA M, MIMOUNEB SM. Modified Jiles–Atherton model and parameters identification using false position method [J]. Physica B: Condensed Matter, 2010,405(8):1947-1950.

[24] TRAPANESE, MARCO. Identification of parameters of the Jiles-Atherton model by neural networks [J]. Journal of Applied Physics, 2011(109):07D355.

[25] GYORGY K. On the product Preisach model of hysteresis [J]. Physica B: Condensed Matter, 2000,275(1-3):40-44.

[26] WANG X, REYSETT A, POMMIER BV, et al. A modified Preisach model and its inversion for hysteresis compensation in piezoelectric actuators [J]. Multidiscipline Modeling in Materials and Structures, 2014,10(1):122-142.

[27] MAYERGOYZ I, CLAUDIO S. Nonlinear diffusion and the Preisach model of hysteresis [J]. Physica B: Condensed Matter. 2000,275(1-3):17-23.

[28] SPANOS PD, CACCIOLA P, REDHORSE J. Random Vibration of SMA Systems via Preisach Formalism [J]. Nonlinear Dynamics, 2004,36(2-4):405-419.

[29] CROSS R, KRASNOSEL'SKII AM, POKROVSKII1 AV. A time-dependent Preisach model [J]. Physica B: Condensed Matter. 2001,306(1-4):206-210.

[30] 高学辉，任雪梅，巩兴安，等．基于分段一致方法的迟滞Preisach模型辨识[C]．第32届中国控制会议（CCC2013），西安，2013:1680-1685.

[31] GAO XH, REN XM, ZHU CS, et al. Discrete composite control for piezoelectric actuator systems [C]. Control and Decision Conference (CCDC), 2015 27th Chinese, 2015, 4469-4473.

[32] GAO XH, REN XM, ZHU CS, et al. Identification and control for Hammerstein systems with hysteresis non-linearity [J]. IET Control Theory and Applications, 2015, 9(13): 1935-1947.

[33] CHEN YS, QIU JH, PALACIOS J, et al. Tracking control of piezoelectric stack actuator using modified Prandtl–Ishlinskii model [J]. Journal of Intelligent Material Systems and Structures, 2013, 24(6):753-760.

[34] MOHAMMAD AJ, PAVEL K. Prandtl–Ishlinskii hysteresis models for complex time dependent hysteresis nonlinearities [J]. Physica B: Condensed Matter, 2012, 407(9):1365-1367.

[35] HUA CC, LI YF. Output feedback prescribed performance control for interconnected time-delay systems with unknown Prandtl–Ishlinskii hysteresis [J]. Journal of the Franklin Institute, 2015, 352(7):2750-2764.

[36] ZHANG J, EMMANUELLE M, NELSON S, et al. Optimal compression of generalized Prandtl–Ishlinskii hysteresis models [J]. Automatica, 2015, 57:170-179.

[37] OMAR A, MOHAMMAD AJ, SUBHASH R, et al. Compensation of rate-dependent hysteresis nonlinearities in a magnetostrictive actuator using an inverse Prandtl–Ishlinskii model [J]. Smart Materials and Structures, 2013, 22(2):025027.

[38] MOHAMMAD AJ, SUBHASH R, SU CY. A generalized Prandtl–Ishlinskii model for characterizing the hysteresis and saturation nonlinearities of smart actuators [J]. Smart Materials and Structures, 2009,18(4):045001.

[39] LIN CJ, LIN CR, YU SK, et al. Hysteresis modeling and tracking control for a dual pneumatic artificial muscle system using Prandtl–Ishlinskii model [J]. Mechatronics, 2015,28:35-45.

[40] YANG MJ, GU GY, ZHU LM. Parameter identification of the generalized Prandtl–Ishlinskii model for piezoelectric actuators using modified particle swarm optimization [J]. Sensors and Actuators A: Physical, 2013(189):254-265.

[41] ZHU W, WANG DH. Non-symmetrical Bouc–Wen model for piezoelectric ceramic actuators [J]. Sensors and Actuators A: Physical, 2012(181):51-60.

[42] SIRETEANU T, GIUCLEA M, MITU A, et al. A Genetic Algorithms Method for Fitting the Generalized Bouc-Wen Model to Experimental Asymmetric Hysteretic Loops [J]. Journal of Vibration and Acoustics, Transactions of the ASME, 2012,134(4):041007.

[43] FAYCAL I, VÍCTOR M, RODELLAR J. Dynamic proper-ties of the hysteretic Bouc-Wen model [J]. Systems and Control Letters, 2007,56(3):197-205.

[44] GILBERTO AO, DIEGO AA, DANIEL BR. Identification of Bouc–Wen type models using multi-objective optimization algorithms [J]. Computers and Structures, 2013:121-132.

[45] XUE XM, WU XH, CHEN LQ, et al. Bouc-Wen modeling to hysteresis nonlinear in Macro Fiber Composite (MFC) actuator [J]. International Journal of Applied Electromagnetics and Mechanics, 2014,45(1-4):965-971.

[46] JILES DC. ATHERTON DL. Theory of ferromagnetic hysteresis [J]. Journal of Applied Physics, 1984(55):2115-2120.

[47] RAMESH A, JILES DC, RODERICK JM. A model of anisotropic a hysteretic magnetization [J]. IEEE Transactions on Magnetics, 1996,32(5):4234-4236.

[48] PREISACH F. über die magnetische Nachwirkung [J]. Zeitschrift für Physik(in German),1935(94):277-302.

[49] AUGUSTO V. Differential Models of Hysteresis [M]. Berlin: Springer Verlag. 1994.

[50] WEISS P, FREUDENREICH J. Etude de l' aimantation initiale en fonction de latemp' erature [J]. Archives des Scienes Physiques et Naturelles((Gen 'eve) in French),1916(42):449-470.

[51] KRASNOSEL, SKII MA, POKROVSKII AV. Systems with Hysteresis [M]. Berlin-Heidelberg-New York-Paris-Tokyo: Springer Verlag. 1989.

[52] PRANDTL L. Ein Gedankenmodell zur kinetischen Theorie der festen Korper [J]. Journalof Applied Mathematics and Mechanics / Zeitschrift für Angewandte Mathematik und Mechanik(in German), 1928,8(2):85-106.

[53] ISHLINSKII AY. On the plane motion of sand [J]. Ukrain. Matern. Zhurnal (in Russian).1954,6(4):430-441.

[54] BOUC R. Modèle mathématique d'hystérésis: application aux systèmes àun degréde liberté [J]. Acustica (in French), 1971(24):16-25.

[55] WEN YK. Method for random vibration of hysteretic systems [J]. Journal of Engineering Mechanics, 1976,102(2):249-263.

[56] JI W, LI Q, XU B, et al. Adaptive fuzzy PID composite control with hysteresis band switching for line of sight stabilization servo system [J]. Aerospace Science and Technology, 2011,15(1):25-32.

[57] TUNAY I, KAYNAK O. Provident control of an electro-hydraulic servo with experimental results [J]. Mechatronics, 1996,6(3):249-260.

[58] RAHMAN MA, ABDULLAH AM, YAO K. Analysis and modeling of hysteresis of piezoelectric micro-actuator used in high precision dual-stage servo system [J]. Control Theory and Technology, 2015,13(2):184-203.

[59] CHAO PAUL CP, LIAO PY, TSAI MY, et al. Robust control design for precision positioning of a generic piezoelectric system with consideration of microscopic hysteresis effects [J]. Micro-system Technologies, 2011,17(5-7):1009-1023.

[60] HARTMUT J, KLAUS K. Real-time compensation of hysteresis and creep in piezoelectric actuators [J]. Sensors and Actuators A: Physical, 2000,79(2):83-89.

[61] MA LW, SHEN Y, LI JR, et al. A modified HO-based model of hysteresis in piezoelectric actuators [J]. Sensors and Actuators A: Physical,2014(220):316-322.

[62] MOHAMMAD RZ, HASSAN S. Experimental comparison of some phenomenological hysteresis models in characterizing hysteresis behavior of shape memory alloy actuators [J]. Journal of Intelligent Material Systems and Structures,2012,23(12):1287-1309.

[63] ASUA E, ETXEBARRIA V, ARRIBAS AG. Neural network-based micro-positioning control of smart shape memory alloy actuators [J]. Engineering Applications of Artificial Intelligence, 2008,21(5):796-804.

[64] AHRENS JH, TAN XB, KHALIL HK. Multirate Sampled-Data Output Feedback Control with Application to Smart Material Actuated Systems [J]. IEEE Transactions on Automatic Control,2009,54(11):2518-2529.

[65] ZISKE J, EHLE F, NEUBERT H, et al. A Simple Phenomenological Model for Magnetic Shape Memory Actuators [J]. IEEE Transactions on Magnetics,2015,51(1):4002608.

[66] VRIJSEN NH, JANSEN JW, COMPTER JC, et al. Measurement method for determining the magnetic hysteresis effects of reluctance actuators by evaluation of the force and flux variation [J]. Review of Scientific Instruments,2013(84):075003.

[67] DIMITROPOULOS PD, STAMOULIS GI, HRISTOFOROU E. A 3-D hybrid Jiles-Atherton/Stoner-Wohlfarth magnetic hysteresis model for inductive sensors and actuators [J]. IEEE Sensors Journal, 2006,6(3):721-736.

[68] STEINER H, STIFTER M, HORTSCHITZ W, et al. Planar Magnetostrictive

Micromechanical Actuator [J]. IEEE Transactions on Magnetics, 2015,51(1):4700104.

[69] LI WJ, YADMELLAT P, KERMANI MR. Linearized Torque Actuation Using FPGA-Controlled Magnetorheological Actuators [J]. IEEE/ASME Transactions on Mechatronics, 2015,20(2):696-704.

[70] VRIJSEN NH, JANSEN JW, LOMONOVA EA. Prediction of Magnetic Hysteresis in the Force of a Prebiased E-Core Reluctance Actuator [J]. IEEE Transactions on Industry Applications, 2014,50(4):2476-2484.

[71] REN XM, LV XH. Identification of Extended Hammerstein Systems Using Dynamic Self-Optimizing Neural Networks [J]. IEEE Transactions on Neural Networks, 2011,22(8):1169-1179.

[72] SKRJANC I, BLAZIC S, AGAMENNONI OE. Interval fuzzy modeling applied to Wiener models with uncertainties [J]. IEEE Transactions on Systems, Man, and Cybernetics, Part B: Cybernetics, 2005,35(5):1092-1095.

[73] TAN AH, GODFREY K. Modeling of direction-dependent Processes using Wiener models and neural networks with nonlinear output error structure [J]. IEEE Transactions on Instrumentation and Measurement, 2004,53(3):744-753.

[74] TAN AH, GODFREY K. Identification of Wiener-Hammerstein models using linear interpolation in the frequency domain (LIFRED) [J]. IEEE Transactions on Instrumentation and Measurement,2002,51(3):509-521.

[75] NI B, GILSON M, GARNIER H. Refined instrumental variable method for

Hammerstein-Wiener continuous-time model identification [J]. IET Control Theory and Applications, 2013,7(9):1276-1286.

[76] TAN YH, DONG RL, LI RY. Recursive Identification of Sandwich Systems with Dead Zone and Application [J]. IEEE Transactions on Control Systems Technology, 2009,17(4):945-951.

[77] TAN XB, JOHN SB. Adaptive Identification and Control of Hysteresis in Smart Materials [J]. IEEE Transactions on Automatic Control, 2005,50(6):827-839.

[78] FALLAH E, MOGHANI JS. A New Identification and -Procedure for the Isotropic Vector Preisach Model [J]. IEEE Transactions on Magnetics,2008,44(1):37-42.

[79] RUDERMAN M, BERTRAM T. Identification of Soft Magnetic B-H Characteristics Using Discrete Dynamic Preisach Model and Single Measured Hysteresis Loop [J]. IEEE Transactions on Magnetics, 2012,48(4):1281-1284.

[80] GAO XH, REN XM, ZHU CS, et al. Identification and control for hammerstein systems with hysteresis non-linearity [J]. IET Control Theory Appl, 2015, 9(13):1935–1947.

[81] SONG G, ZHAO JQ, ZHOU XQ, et al. Tracking control of a piezoceramic actuator with hysteresis compensation using inverse Preisach model [J]. IEEE/ASME Transactions on Mechatronics, 2005,10(2):198-209.

[82] BAO KN, KYOUNG KA. Feedforward Control of Shape Memory Alloy Actuators Using Fuzzy-Based Inverse Preisach Model [J]. IEEE Transactions on Control Systems Technology, 2009,17(2):434-441.

[83] LIU L, TAN KK, CHEN SL, et al. Discrete Composite Control of Piezoelectric Actuators for High-Speed and Precision Scanning [J]. IEEE Transactions on Industrial Informatics, 2013,9(2):859-868.

[84] GAO XH, REN XM, ZHANG CY, et al. Online identification and robust adaptive control for discrete hysteresis preisach model [C]. Lecture Notes in Electrical Engineering (CISC2017), 2016(359): 41-49.

[85] XIAO SL, LI YM. Modeling and High Dynamic Compensating the Rate-Dependent Hysteresis of Piezoelectric Actuators via a Novel Modified Inverse Preisach Model [J]. IEEE Transactions on Control Systems Technology,2013,21(5):1549-1557.

[86] GAO XH, SUN B, WANG SB. Hopfield Neural Network Identification for Preisach Hysteresis System [C]. Chinese Control Conference, CCC2018, 2018: 1580-1584.

[87] WEI TA, KHOSLA PK, RIVIERE CN. Feedforward Controller with Inverse Rate-Dependent Model for Piezoelectric Actuators in Trajectory-Tracking Applications[J]. IEEE/ASME Transactions on Mechatronics, 2007,12(2):134-142.

[88] TAN UX, LATT WT, SHEE CY, et al. Feedforward Controller of Ill-Conditioned Hysteresis Using Singularity-Free Prandtl–Ishlinskii Model [J]. IEEE/ASME Transactions on Mechatronics,2009,14(5):598-605.

[89] CHEN XK，HISAYAMA T, SU CY. Adaptive Control for Uncertain Continuous-Time Systems Using Implicit Inversion of Prandtl-Ishlinskii

Hysteresis Representation [J]. IEEE Transactions on Automatic Control, 2010,55(10):2357-2363.

[90] LIU SN, SU CY, LI Z. Robust Adaptive Inverse Control of a Class of Nonlinear Systems With Prandtl-Ishlinskii Hysteresis Model [J]. IEEE Transactions on Automatic Control, 2014,59(8):2170-2175.

[91] LIU SN, SU CY. Inverse error analysis and adaptive output feedback control of uncertain systems preceded with hysteresis actuators [J]. IET Control Theory and Applications, 2014,8(17):1824-1832.

[92] LI Z, HU Y, LIU Y, et al. Adaptive inverse control of nonlinear systems with unknown complex hysteretic nonlinearities [J]. IET Control Theory and Applications, 2012,6(1):1-7.

[93] GU GY, ZHU LM, SU CY. Modeling and Compensation of Asymmetric Hysteresis Nonlinearity for Piezoceramic Actuators with a Modified Prandtl–Ishlinskii Model [J]. IEEE Transactions on Industrial Electronics,2014,61(3):1583-1595.

[94] JANAIDEH M, KREJCI P. Inverse Rate-Dependent Prandtl–Ishlinskii Model for Feedforward Compensation of Hysteresis in a Piezomicropositioning Actuator [J]. IEEE/ASME Transactions on Mechatronics, 2013,18(5):1498-1507.

[95] QIN YD, TIAN YL, ZHANG DW, et al. A Novel Direct Inverse Modeling Approach for Hysteresis Compensation of Piezoelectric Actuator in Feedforward Applications [J]. IEEE/ASME Transactions on Mechatroni

cs,2013,18(3):981-989.

[96] SU CY, WANG QQ, CHEN XK, et al. Adaptive variable structure control of a class of nonlinear systems with unknown Prandtl-Ishlinskii hysteresis [J]. IEEE Transactions on Automatic Control, 2005,50(12):2069-2074.

[97] CHEN XK, SU CY, FUKUDA T. Adaptive Control for the Systems Preceded by Hysteresis [J]. IEEE Transactions on Automatic Control, 2008,53(4):1019-1025.

[98] ZHANG XY, LIN Y. Adaptive tracking control for a class of pure-feedback nonlinear systems including actuator hysteresis and dynamic uncertainties [J]. IET Control Theory and Applications, 2011,5(10):1868-1880.

[99] REN BB, GE SS, SU CY, et al. Adaptive Neural Control for a Class of Uncertain Nonlinear Systems in Pure-Feedback Form With Hysteresis Input [J]. IEEE Transactions on Systems, Man, and Cyber-netics, Part B: Cybernetics, 2009,39(2):431-443.

[100] REN BB, GE SS, LEE TH, et al. Adaptive Neural Control for a Class of Nonlinear Systems With Uncertain Hysteresis Inputs and Time-Varying State Delays [J]. IEEE Transactions on Neural Networks, 2009,20(7):1148-1164.

[101] GAO XH, WANG SB, LIU RG, et al. Hopfield neural network identification for Prandtl-Ishlinskii hysteresis nonlinear system [C]. Lecture Notes in Electrical Engineering (CISC2018), 2018(528):153-161.

[102] RICCARDI L, NASO D, TURCHIANO B, et al. Adaptive Control of

Positioning Systems With Hysteresis Based on Magnetic Shape Memory Alloys [J]. IEEE Transactions on Control Systems Technology, 2013,21(6):2011-2013.

[103] ASCHEMANN H, SCHINDELE D. Comparison of Model-Based Approaches to the Compensation of Hysteresis in the Force Characteristic of Pneumatic Muscles [J]. IEEE Transactions on Industrial Electronics, 2014,61(7):3620-3629.

[104] BASHASH S, JALILI N. Robust Multiple Frequency Trajectory Tracking Control of Piezoelectrically Driven Micro/Nanopositioning Systems [J]. IEEE Transactions on Control Systems Technology, 2007,15(5):867-878.

[105] MOHAMMED I, FAYCAL I, JOSE R. The Hysteresis Bouc-Wen Model, a Survey [J]. Archives of Computational Methods in Engineering,2009,16(2):161-188.

[106] XU YH, AHMADIAN M, SUN RY. Improving Vehicle Lateral Stability Based on Variable Stiffness and Damping Suspension System via MR Damper [J]. IEEE Transactions on Vehicular Technology, 2014,63(3):1071-1078.

[107] HABINEZA D, RAKOTONDRABE M, LE GY. Bouc–Wen Modeling and Feedforward Control of Multivariable Hysteresis in Piezoelectric Systems: Application to a 3-DoF Piezo tube Scanner [J]. IEEE Transactions on Control Systems Technology,2015,23(5):1797-1806.

[108] LIU Z, LAI GY, ZHANG Y, et al. Adaptive Neural Output Feedback Control of Output-Constrained Nonlinear Systems with Unknown Output

Nonlinearity [J]. IEEE Transactions on Neural Networks and Learning Systems,2015,26(8):1789-1802.

[109] GAO XH. Adaptive neural control for hysteresis motor driving servo system with Bouc-wen model [J]. Complexity,2018(1):1-9.

[110] SU CY, STEPANENKO Y, SVOBODA J, et al. Robust adaptive control of a class of nonlinear systems with unknown backlash-like hysteresis [J]. IEEE Transactions on Automatic Control, 2000,45(12):2427-2432.

[111] SU CY, OYA M, HONG H. Stable adaptive fuzzy control of nonlinear systems preceded by unknown backlash-like hysteresis [J]. IEEE Transactions on Fuzzy Systems, 2003,11(1):1-8.

[112] ZHOU J, WEN CY, ZHANG Y. Adaptive backstepping control of a class of uncertain nonlinear systems with unknown backlash-like hysteresis [J]. IEEE Transactions on Automatic Control, 2004,49(10):1751-1759.

[113] WANG HQ, CHEN B, LIU KF, et al. Adaptive Neural Tracking Control for a Class of Nonstrict-Feedback Stochastic Nonlinear Systems with Unknown Backlash-Like Hysteresis [J]. IEEE Transactions on Neural Networks and Learning Systems, 2014,25(5):947-958.

[114] HUANG XY, ZHANG H, ZHANG GG, et al. Robust Weighted Gain-Scheduling H1 Vehicle Lateral Motion Control with Considerations of Steering System Backlash-Type Hysteresis [J]. IEEE Transactions on Control Systems Technology, 2014,22(5):1740-1753.

［115］LIU Y, LIN Y. Global adaptive output feedback tracking for a class of non-linear systems with unknown backlash-like hysteresis［J］. IET Control Theory and Applications, 2014,8(11):927-936.

［116］LIU RG, GAO XH. Neural network identification and sliding mode control for hysteresis nonlinear system with backlash-like model［J］. Complexity, 2019(1):1-9.

第 2 章 Preisach 迟滞非线性系统下三角矩阵分段一致辨识与滑模控制

2.1 问题的提出

多个学者针对迟滞系统的辨识研究做了大量工作，但是由于迟滞非线性的多值性和记忆性以及频率相关性，使得迟滞非线性的辨识一直是研究的难点问题。迟滞非线性环节的多值性，大大增加了迟滞系统的辨识难度；迟滞非线性环节的记忆性和频率相关性，又进一步提高了辨识的要求。虽然学者们做了大量工作，也取得了一些好的结果，比如应用基于最小二乘法的辨识等，但是受限于迟滞模型研究和迟滞系统本身存在的复杂性，使得迟滞模型辨识仍然是迟滞系统控制中的难点问题。本章针对用 Preisach 迟滞模型描述的 Hammerstein 系统的迟滞环节辨识问题进行研究，找出新的辨识方法，应用矩阵计算，辨识 Preisach 迟滞密度函数，完成 Preisach 迟滞系统的辨识与控制。

Preisach 模型的辨识从 20 世纪 70 年代开始至今都是研究的热点问题。梅尔戈兹[1]提出的一阶转换曲线辨识法一直广为引用，但是随着技术的进步和对模型精确性要求的提高,该方法受 Preisach 模型的"擦除"特性影响的问题越来越难以回避。谭小波等学者[2]提出了两类最小二乘递归辨识法辨识 Preisach 模型,郭咏新等学者[3]应用最小二乘法辨识 Preisach 模型均可消除"擦除"特性的影响。不仅如此,一些学者采用智能方法[4-7]也可不受"擦除"特性影响的辨识 Preisach 模型。李春涛和谭永红[5]采用神经网络模型逼近 Preisach 模型实现对迟滞模型的辨识,而阮保嘉和安景宽[8]则用模糊推理系统估计 Preisach 迟滞模型。即使如此,上述方法要么采用最小二乘递归计算,要么采用智能计算,虽然消除了 Preisach 模型"擦除"特性的影响,但是其计算量相对梅尔戈兹[1]提出的经典辨识方法要复杂。如何在消除"擦除"特性影响的基础上降低运算复杂度一直是未完成的问题。

2.2 研究内容

Hammerstein 系统是指非线性环节串联线性系统组成的非线性系统。对于 Hammerstein 系统的辨识,学者们从不同角度进行了深入研究[9-12],取得了一些不错的结果。李国旗和温常云[11]研究了 Hammerstein 系统的定点迭代辨识法并讨论了任意非零条件下的一致估计。而达菲拉和韦斯特威克[13]采用一种扩展的支持向量机辨识由各态历经自回归模型描述的 Hammerstein 系统。任雪梅和吕晓华[14]则采用一种新的最优化方法针对非高斯噪声的扩展 Hammerstein 系统实行动态自最优神经网络辨识与控制。与文献[14]不同,米查尔基维茨[15]采用改进的科尔莫戈洛夫神经网络辨识 Hammerstein 系统。但是尚无学者针对迟滞 Hammerstein 系统的辨识进行盲辨识研究。

盲辨识法由白尔维等学者[9,16]提出,最早应用于通信系统,通过过采样输出将单输入单输出（single input single output，SISO）系统转化为单输入多输出（single input multi-output，SIMO）系统来辨识系统模型。文献［16］首先证明了一个 n 阶系统可盲辨识的充要条件是过采样率 p 满足 $p \geqslant n+1$。白尔维和傅敏越[9]研究了系统阶次 n 已知情况下的 Hammerstein 系统的盲辨识，王建东等学者[17]在文献［9］的基础上进一步研究，将其扩展到闭环系统。而俞成浦等学者[10]进一步放松条件，提出了一种确定性盲辨识算法，该算法的过采样率小于分子多项式长度。但是以上研究均没有特别针对迟滞 Hammerstein 系统且均需系统阶次 n 已知。

  本章放松了盲辨识的条件，在不需要模型阶次 n 已知的情况下，仅仅依靠输入输出数据辨识用 Preisach 模型描述的迟滞 Hammerstein 系统，提出了新的下三角矩阵法辨识 Preisach 模型，该方法不但避免了"擦除"特性的影响，而且采用矩阵运算减少了运算复杂度。在辨识结果的基础上设计了逆模型和滑模控制结合的混合控制策略对迟滞 Hammerstein 系统进行精确控制，该控制方案不仅提高了收敛速度并且提高了控制系统的鲁棒性。主要创新总结如下：

  （1）提出了一种新的确定性盲辨识方法辨识迟滞 Hammerstein 系统，该方法放松了盲辨识的条件，仅仅需要输入输出数据而不需要系统阶次已知。然后将 Preisach 模型从积分形式转化为矩阵形式，从而使辨识 Preisach 函数问题转化为辨识 Preisach 系数矩阵问题，通过三角矩阵计算辨识 Preisach 密度函数。

  （2）设计了一种新的混合控制策略，由离散逆模型控制器（discrete inverse model-based control，DIMBC）和离散自适应滑模控制器（discrete adaptive sliding mode control，DASMC）两部分组成。所提出的控制策略通过离散逆模型控制器减少控制系统收敛时间，通过离散自适应滑模控制器提高控制系统的鲁棒性且应用自适应参数降低滑模控制器的抖震。

2.3 问题描述

Hammerstein 系统线性环节离散形式可用如下传递函数表示：

$$G(z)=\frac{y(z)}{\upsilon(z)}=\frac{b_1 z^{-1}+b_2 z^{-2}+\cdots+b_n z^{-n}}{1+a_1 z^{-1}+a_2 z^{-2}+\cdots+a_n z^{-n}} \quad (2.1)$$

其中，n 表示系统阶次，$a_i, b_i, =1, 2, \cdots, n$ 为 $G(z)$ 的未知系数。Hammerstein 系统非线性环节 $v(t)=f(u(t))$ 在本章中用 Preisach 模型描述，表示迟滞非线性。

Preisach 模型由 Preisach 算子组成，Preisach 算子定义为[2]：考虑一对阈值 α，β 且 $\alpha > \beta$，则 Preisach 算子 $\hat{\gamma}_{\alpha,\beta}[\bullet,\bullet]$ 可表示为

$$\hat{\gamma}_{\alpha,\beta}[u,\varsigma] \triangleq \begin{cases} 0, & u(t) < \beta \\ 1, & u(t) > \alpha \\ \hat{\gamma}_{\alpha,\beta}[u,\varsigma](t^-), & \beta \leq u(t) \leq \alpha \end{cases} \quad (2.2)$$

其中，$u \in \mathbb{C}([0,T])$，$\varsigma \in \{1,0\}$ 或 $\{1,-1\}$，$t \in [0,T]$，$\hat{\gamma}_{\alpha,\beta}[u,\varsigma](0^{-1})=\varsigma$ 且 $t^{-1}=\lim_{\varepsilon<0, \varepsilon\to 0} t \to \varepsilon$。

定义 Preisach 平面 $\mathcal{P}_0$：

$$\mathcal{P}_0 \triangleq \{(\alpha,\beta) \in \mathbb{R}^2 : \alpha \geq \beta\} \quad (2.3)$$

则任意 $(\alpha,\beta) \in \mathcal{P}_0$ 受 $\hat{\gamma}_{\alpha,\beta}$ 限制。若 $u \in \mathbb{C}([0,T])$ 且符合波莱尔可测初始条件 $\varsigma_0: \mathcal{P}_0 \to \{1,0\}$，则 Preisach 模型可表示为[2]

$$\upsilon(t)=f(u(t))=\int_{\mathcal{P}_0} \mu(\alpha,\beta) \hat{\gamma}_{\alpha,\beta}[u,\varsigma_0(\alpha,\beta)](t) \mathrm{d}\alpha \mathrm{d}\beta \quad (2.4)$$

其中，μ 为未知 Preisach 密度函数。实际情况下，μ 为紧集，即对 α_0，β_0，若 α

$>\alpha_0$ 或 $\hat{a}<\beta_0$，有 $\mu(\alpha,\beta)=0$。因此，Preisach模型的输出满足在Preisach平面 $\mathcal{P} \triangleq \{(\alpha,\beta)\in\mathcal{P}_0 | \alpha \leq \alpha_0, \beta \geq \beta_0\}$ 上为有限的三角区域。

本章的目标为：

（1）估计 Hammerstein 系统线性环节式（2.1）的阶次 n；

（2）辨识式 (2.1) 的未知参数 $\{a_1, a_2, \cdots, a_n\}$，$\{b_1, b_2, \cdots, b_n\}$；

（3）辨识 Hammerstein 系统非线性环节 $v(t)=f(u(t))$，或者进一步辨识 Preisach 模型的未知 Preisach 密度函数 μ；

（4）设计由离散逆模型控制和离散自适应滑模控制组成的混合控制策略精确控制迟滞 Hammerstein 系统。

2.4 Hammerstein 系统辨识

大多数 Hammerstein 系统辨识方法是在系统阶次 n 已知的前提下进行的，而本章所讨论的迟滞 Hammerstein 系统辨识方法假设系统阶次 n 未知。因此，为了辨识 Hammerstein 系统，需要首先获得 Hammerstein 系统阶次 n。由于仅仅输入输出数据已知，与文献［18-19］不同，本章采用 Hankel 矩阵法估计 Hammerstein 系统的阶次，此方法仅仅需要 Hammerstein 系统的输出数据 $y(t)$。

当系统阶次 n 确定以后，即式（2.1）中 n 已经确定，则可根据过采样的输出数据 $y(t)$ 通过盲辨识法辨识传递函数 $G(z)$；由 $G(z)$ 可计算出 Hammerstein 系统线性环节输入 $v(t)$，此即 Hammerstein 系统非线性环节输出；又已知 Hammerstein 系统输入 $u(t)$，根据本章提出的三角矩阵辨识法辨识 Preisach 迟滞模型的 Preisach 密度函数从而得到 Preisach 模型的辨识结果完成 Hammerstein 系统的辨识。

2.4.1 估计系统阶次 n

系统输入设定为脉冲序列，输出的脉冲响应为 $y(1)$，$y(2)$，…，$y(L)$。

那么可定义 Hankel 矩阵：

$$H(l,j) = \begin{bmatrix} y(l) & y(l+1) & \cdots & y(l+j-1) \\ y(l+1) & y(l+2) & \cdots & y(l+j) \\ \cdots & \cdots & & \cdots \\ y(l+j-1) & y(l+j) & \cdots & y(l+2j-2) \end{bmatrix} \quad (2.5)$$

其中，j 表示Hankel矩阵维数，l决定用哪些脉冲响应序列组成Hankel矩阵满足 $l \in [1, L-2j+2]$。

对 $j \in [1,L]$，计算 Hankel 矩阵行列式 $\det[H(l,j)]$，当 $\det[H(l,j)]=0$ 时，令 $j=n$，即为系统阶次。但是在实际应用中 Hammerstein 系统往往受到弱噪声的影响，很难求得 $\det[H(l,j)]=0$，即当 $j=n$ 时，$\det[H(l,j)] \neq 0$。在这种情况下，定义 Hankel 矩阵的平均行列式[20]如下：

$$D = \arg\max_{1 \leq j \leq L} \frac{\frac{1}{L-2j+2}\sum_{l=1}^{L-2j+2}\det[H(l,j)]}{\frac{1}{L-2j}\sum_{l=1}^{L-2j}\det[H(l,(j+1))]} \quad (2.6)$$

根据式 (2.6)，当 $j \in [1,L]$ 时，计算 D，虽然由于弱噪声的影响，即使 $j=n$ 时，D 的分母 $\frac{1}{L-2j}\sum_{l=1}^{L-2j}\det[H(l,(j+1))] \neq 0$，但是与分子 $\frac{1}{L-2j+2}\sum_{l=1}^{L-2j+2}\det[H(l,j)]$ 相比将快速减小，即当 D 取得最大值时，j 就是系统阶次 n，得到 $G(z)$ 的阶次。

2.4.2 $G(z)$ 分母的估计

确定系统阶次 n 以后，用盲辨识算法辨识 $G(z)$。将输入采样间隔定为 T，输

出采样间隔可定为 $h=\dfrac{T}{\rho}, \rho \geqslant 1$. 根据文献 [16]，$n$ 阶系统可盲辨识的充要条件为 $\rho \geqslant n+1$。为了简化设计，在本章中选取 $\rho=n+1$，根据式（2.1），下式成立[9]：

$$G_{n+1}(z)=\frac{y_{n+1}(z)}{x_{n+1}(z)}=\frac{\bar{b}_1 z^{-1}+\bar{b}_2 z^{-2}+\cdots+\bar{b}_n z^{-n}}{1+\bar{a}_1 z^{-1}+\bar{a}_2 z^{-2}+\cdots+\bar{a}_n z^{-n}} \tag{2.7}$$

进一步将 (2.7) 改写为参数化形式：

$$y[(k+1)h]=F^T[kh]\phi \tag{2.8}$$

其中，$\begin{cases} F[kh]=\left[-y[(k-1)h],\cdots,-y[(k-n)h],\upsilon[(k-1)h],\cdots,\upsilon[(k-n)h]\right]^T \\ \phi \quad =\left[\bar{a}_1,\bar{a}_2,\cdots,\bar{a}_n,\bar{b}_1,\bar{b}_2,\cdots,\bar{b}_n\right]^T \end{cases}$

注意到式 (2.8) 中包含不可测变量 $v(t)$，因此其系数无法直接估计。但是输入序列 $v(t)$ 当且仅当 $t=kT=k(n+1)h$ 时不为零，即如果 $l=k(n+1)$，则 $\upsilon[(l-1)h]=\upsilon[(l-2)h]$ 成立。在这种情况下，下式成立：

$$y[kT]=\bar{F}^T[h]\bar{\phi} \tag{2.9}$$

其中，$\begin{cases} \bar{F}[h]=\left[-y[kT-h],-y[kT-2h],\cdots,-y[kT-nh]\right]^T \\ \bar{\phi} \quad =\left[\bar{a}_1,\bar{a}_2,\cdots,\bar{a}_n\right]^T \end{cases}$

显然，式 (2.9) 仅仅包含可测的输出，因此可估计其参数 $\bar{\phi}$。$\bar{\phi}$ 可以有多种估计方法，本书采用递归加权最小二乘法来估计。对每一个 k，最小二乘算法给出如下：

$$\begin{cases} \hat{\bar{\phi}}(k)=\hat{\bar{\phi}}(k-1)+K(k)\left[y(k)-\bar{F}^T(k)\hat{\bar{\phi}}(k-1)\right] \\ K(k)=P(k-1)\bar{F}(k)\left[\bar{F}^T(k)P(k-1)\bar{F}(k)+\dfrac{1}{\lambda(k)}\right]^{-1} \\ P(k)=\left[I-K(k)\bar{F}^T(k)\right]P(k-1) \end{cases} \tag{2.10}$$

其中，$\lambda>0$ 为最小二乘法的权值。

2.4.3 $G(z)$ 分子的估计

2.4.2 节讨论了 $G(z)$ 分母的辨识，本节将估计分子的系数。首先考虑一下两个输出序列 $Y_{kT}(z) = \sum_{k=1}^{\infty} y[kT]z^{-k} = G_{kT}(z)X_{kT}(z), Y_{kT-h}(z) = \sum_{k=1}^{\infty} y[kT-h]z^{-k} = G_{kT-h}(z)X_{kT-h}(z)$ 在采样间隔 $h = \dfrac{T}{n+1}$ 时，有 $X_{kT}(z) = \sum_{k=0}^{\infty} \upsilon[kT] = X_{kT-h}(z)$，则 $G_{kT-h}(z)$ 可以表示为 $G_{kT-h}(z) = \dfrac{b_0 + b_1 z^{-1} + \cdots + b_{n-1} z^{-1}}{1 + \overline{a}_1 z^{-1} + \overline{a}_2 z^{-2} + \cdots + \overline{a}_n z^{-n}}$。

因此，$G_{kT}(z)$ 和 $G_{kT-h}(z)$ 具有相同的分母，可得 $G_{kT-h}(z)X_{kT-h}(z) - G_{kT}(z)X_{kT}(z) = 0$。这表示：

$$\overline{\theta}_{kT-h}(z)Y_{kT}(z) - \overline{\theta}_{kT}(z)Y_{kT-h}(z) = 0 \tag{2.11}$$

其中，$\overline{\theta}_{kT}(z) = [\overline{b}_1, \overline{b}_2, \cdots, \overline{b}_n]^T$，$\overline{\theta}_{kT-h}(z) = [b_0, b_1, \cdots, b_{n-1}]^T$。为了便于辨识 $G(z)$ 的分母，将式（2.11）改成为：

$$y[kT] = E^T(h)\varphi \tag{2.12}$$

其中，

$$\begin{cases} E(h) = \left[-y[kT-h], \cdots, -y[kT-(n-1)h], \cdots, y[kT-(n+1)h]\right]^T \\ \varphi = \dfrac{1}{b_0}\left[\overline{b}_1, \overline{b}_2, \cdots, \overline{b}_n, b_1, b_2, \cdots, b_{n-1}\right]^T \end{cases}$$

对每个 k，假设 $b_0 \neq 0$，则 φ 可以用最小二乘法估计：

$$\begin{cases} \hat{\varphi}(k) = \varphi(k-1) + K(k)\left[y(k) - E^T(k)\varphi(k-1)\right] \\ K(k) = P(k-1)E(k)\left[E^T(k)P(k-1)E(k) + \dfrac{1}{\lambda(k)}\right]^{-1} \\ P(k) = \left[I - K(k)E^T(k)\right]P(k-1) \end{cases} \tag{2.13}$$

其中，$\lambda>0$ 为最小二乘法的权值。

注意到 b_i 是一个标量，可以规范化为 $b_0=1$[21]，正则化后的参数可以通过式（2.13）辨识得到。

2.4.4　Preisach 迟滞模型的估计

Hammerstein 系统线性环节传递函数 $G(z)$ 被辨识以后，无法直接测量的线性环节输入变量 $v(t)$ 在采样间隔 T 下可通过已经辨识的式（2.1）和可测量的输出 $y(t)$ 计算得到。而 $v(t)$ 同时为 Hammerstein 系统非线性环节输出变量，对 Preisach 迟滞非线性模型，其输入 $u(t)$ 可测量，输出 $v(t)$ 已经辨识得到，本节将提出新的确定性基于三角矩阵的辨识算法辨识 Preisach 迟滞模型。

经典的 Preisach 模型辨识算法是通过阶梯状曲线将 Preisach 平面分为两个子平面，例如一个典型迟滞曲线如图 2.1 所示，经典辨识算法的 Preisach 平面如图 2.2 所示。迟滞上升曲线（图 2.1 中 a，c 段）在图 2.2 中表示为水平线（图 2.2 中 a，c 段），迟滞下降曲线（图 2.1 中 b，d 段）在图 2.2 中表示为垂直线（图 2.2 中 b，d 段）。

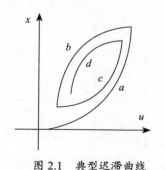

图 2.1　典型迟滞曲线　　　　图 2.2　Preisach 迟滞平面

但是经典辨识方法容易受"擦除"特性影响，即阶梯状曲线可能被"擦除"（关于"擦除"特性详细论述及证明见文献[1]）。为了消除"擦除"特性影响，本节将经典阶梯状曲线划分的 Preisach 平面改进为分离平面，每一段单调曲线独立一个 Preisach 子平面，即 $P_i(1 \leqslant i \leqslant L)$，$L$ 表示单调曲线数（如图 2.1 例子的分离 Preisach 平面见图 2.3，其中 $L=4$）。因此，每一个子平面代表一段单调曲线（单调上升或单调下降）。若第一个子平面表示单调上升（图 2.3（a），表示图 2.1 中上升段 a），则所有奇数子平面均表示单调上升段（图 2.3（a）（c）），偶数子平面表示单调下降段（图 2.3（b）（d）），反之亦然。所以不同单调曲线表示为不同 Preisach 子平面，保留了 Preisach 迟滞曲线的所有细节，从而避免了"擦除"特性的影响。

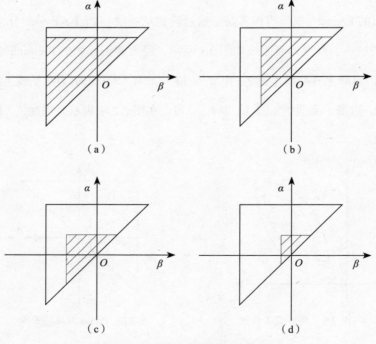

图 2.3 分段一致 Preisach 迟滞平面

令 $[u_{\min}, u_{\max}]$ 为 Hammerstein 系统的实际输入范围，其通常为 $[\alpha_0, \beta_0]$ 的严格子集。如果 $[u_{\min}, u_{\max}]$ 为 m 维，在瞬时时刻 k，Preisach 密度函数在紧集内可离散化为 m 维函数。当有限输入为 $\{u_i\}_{i=1}^{m}$，离散 Preisach 模型可表示为（如图 2.4 所示）：

$$x(k) = \sum_{i=1}^{m} \mu_i(k) \hat{\gamma}_{\alpha,\beta} [u_i, \varsigma(\alpha_i, \beta_i)](k) \tag{2.14}$$

其中，$\mu_i(k)$ 为 Preisach 密度函数且 $k \in [0, T]$。

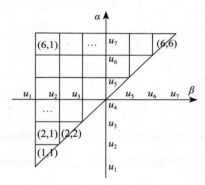

图 2.4 离散 Preisach 平面

当输入为 $\{u_i\}_{i=1}^{m}$ 时，对任意单调上升或下降曲线段，假设输入为 $\{u_j\}_{j=1}^{nL}$，则 $\sum_{i=1}^{L} n_i = m$ 成立。对每一单调曲线段，有如下定理：

定理 2.1 对任意单调一致曲线，其 Preisach 子平面定义为 P_i，如图 2.3 所示，则输入 U 和输出 X 可以表示为 $U = [u_1 \; u_2 \; \ldots \; u_{nL}]^T$，$X = [\upsilon_1 \; \upsilon_2 \ldots \upsilon_{nL}]^T$，将输入 U 和输出 X 扩展为矩阵形式：

$$\hat{U} = \begin{bmatrix} \hat{u}_{11} & 0 & 0 & \cdots & 0 \\ \hat{u}_{21} & \hat{u}_{22} & 0 & \cdots & 0 \\ \hat{u}_{31} & \hat{u}_{32} & \hat{u}_{33} & \cdots & 0 \\ \cdots & \cdots & \cdots & \cdots & \cdots \\ \hat{u}_{nL1} & \hat{u}_{nL2} & \hat{u}_{nL3} & \cdots & \hat{u}_{nLnL} \end{bmatrix}, \; \hat{X} = \begin{bmatrix} \hat{\upsilon}_{11} & 0 & 0 & \cdots & 0 \\ \hat{\upsilon}_{21} & \upsilon_{22} & 0 & \cdots & 0 \\ \hat{\upsilon}_{31} & \upsilon_{32} & \upsilon_{33} & \cdots & 0 \\ \cdots & \cdots & \cdots & \cdots & \cdots \\ \hat{\upsilon}_{nL1} & \upsilon_{nL2} & \upsilon_{nL3} & \cdots & \upsilon_{nLnL} \end{bmatrix} \tag{2.15}$$

其中，$\hat{u}_{ij} = \dfrac{u_i}{i}$，$\hat{v}_{ij} = \dfrac{u_i}{i}$，$j=1, 2, \cdots, i$。则Preisach密度函数$\mu$可以通过下式得到：

$$\mu = \hat{X}U^{-1}\hat{\omega}^{-1} \tag{2.16}$$

且逆Preisach模型可表示为：

$$U = (\mu\hat{\omega})^{-1}X = \delta X \tag{2.17}$$

其中，

$$\hat{\omega} = \begin{bmatrix} 1 & 0 & 0 & \cdots & 0 \\ 1 & 1 & 0 & \cdots & 0 \\ 1 & 1 & 1 & \cdots & 0 \\ \cdots & \cdots & \cdots & & \cdots \\ 1 & 1 & 1 & & 1 \end{bmatrix} \tag{2.18}$$

且 $\delta = (\mu\hat{\omega})^{-1}$。

证明 离散化Preisach平面如图2.4所示，考虑到Preisach算子$\hat{\gamma}_{\alpha,\beta}[u,\zeta(\alpha,\beta)]$的定义和式（2.14），为了简化计算，将图2.4中三角Preisach平面所有单元格中权值设定为"1"，且每一个输入u_i被平均的分配给第i行的单元格。则可将Preisach算子$\hat{\gamma}_{\alpha,\beta}[u,\zeta(\alpha,\beta)]$定义为

$$\bar{\gamma} = \hat{\omega}\hat{U} \tag{2.19}$$

其中，$\bar{\gamma}$表示Preisach算子的任意单调一致曲线段（例如图2.3(a)），$\hat{\omega}$和$\hat{U}$分别在式（2.15）和式（2.18）中定义。

将式（2.14）改写为矩阵形式，同时根据式（2.15），式（2.18）和式（2.19），可以得到$\hat{X}$：

$$\hat{X} = \mu\bar{\gamma} = \mu\hat{\omega}U \tag{2.20}$$

其中，$\hat{U}$和$\hat{\omega}$均为三角阵，则其逆存在且可得，因此有：$\mu = \hat{X}U^{-1}\hat{\omega}^{-1}$。进一步可将式（2.14）改写为

$$X = \mu\hat{\omega}U \quad (2.21)$$

因此，Preisach 逆模型可以表示为

$$U = (\mu\hat{\omega})^{-1}X = \delta X \quad (2.22)$$

证毕。①②③

2.4.5　Hammerstein 系统辨识算法的实现

如前所述，本节所提出的 Hammerstein 系统辨识算法可分为两大步：首先针对 Hammerstein 系统线性环节仅仅依靠输出数据确定系统阶次 n 和传递函数 $G(z)$ 的分子分母未知参数；然后计算出 Hammerstein 系统线性环节输入，此即 Hammerstein 系统非线性环节输出，再根据系统输入辨识出 Preisach 迟滞非线性参数，具体辨识过程如下：

（1）根据式（2.5）和式（2.6）采用 Hankel 矩阵辨识系统阶次 n。

（2）设定输入采样间隔 T，将输出采样间隔设定为 $h = \dfrac{T}{n+1}$。

（3）采集输出数据 $y(t)$。

---

① 式（2.18）中下三角单位阵 $\hat{\omega}$ 是本节所提出的辨识方法的关键，通过它将式（2.14）表示的求和形式转化为矩阵表达形式，简化了计算。从理论上讲，$\hat{\omega}$ 内元素可取任意非零值，本章选取了最简单形式即单位阵的形式，又由于其他矩阵为下三角形式，所以只要保证为 $\hat{\omega}$ 下三角单位阵即可不影响辨识结果。当然也可以选取其他值，这会使得 Preisach 密度函数的辨识结果不同，但不会影响整体结果。

② 定理 2.1 提出了一种基于输入 U 的 n_L 维新的辨识方法，其输入是 m 维系统输入 $\{u_i\}_{i=1}^m$ 的子集。对系统输入 $\{u_i\}_{i=1}^m$ 来说，其每一个子集均可以通过式（2.16）辨识，逆模型可以通过式（2.17）得到，因此整个系统辨识只需 m 维计算即可，小于最小二乘类辨识方法的 $\dfrac{m(m+1)}{2}$ 维计算。

③ 定理 2.1 所提出的辨识 Preisach 密度函数法和求逆模型法均基于 $\varsigma \in \{1,0\}$。对 Preisach 迟滞模型来说，ς 取值范围还有许多种，如 $\varsigma \in \{1,-1\}$ 等。通过坐标变换可以容易地从其他取值范围将 $\varsigma \in \{-1, 1\}$ 变换为 $\varsigma \in \{1,0\}$，从而可以应用定理 2.1 辨识 Preisach 迟滞模型和求逆模型。

（4）采用 Monte-Carlo 法，根据式（2.10）和式（2.13）辨识传递函数 $G(z)$。

（5）根据传递函数 $G(z)$ 和输出数据 $y(t)$，计算线性环节输入（非线性环节输出）$v(t)$。

（6）根据式（2.15）计算 Preisach 迟滞非线性模型每一单调分段的 $\hat{U}$ 和 $\hat{X}$。

（7）根据式（2.16）计算 Preisach 密度函数 μ。

（8）回到第（7）步，辨识下一段单调分段 Preisach 密度函数，直到输入数据 $\{u_i\}_{i=1}^{m}$ 全部完成。①

2.5 混合控制器设计

通过 2.4 节所提出的方法获得 Hammerstein 系统模型以后，即可对该系统进行控制。本节提出了一种混合控制策略控制辨识后的 Hammerstein 系统，该控制策略包括前馈控制器和反馈控制器两个部分，如图 2.5 所示。前馈控制由离散逆模型控制器（discrete inverse model-based controller，DIMBC）构成，分为迟滞非线性环节逆模型和线性环节逆模型两部分；反馈控制由离散自适应滑模控制器（discrete adaptive sliding mode control，DASMC）组成。该控制策略不仅可以通过离散自适应滑模控制器提高控制系统的鲁棒性，而且可以通过离散逆模型控制器提高系统的收敛速度。

---

① 从辨识算法的步骤可以看出，由于下三角矩阵法是确定性辨识算法，因此不影响盲辨识算法的持续激励条件，即盲辨识法的持续激励条件仍然适用于本章提出的 Hammerstein 系统辨识算法。关于盲辨识法持续激励条件的证明可参考文献[9]。

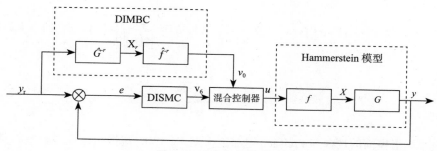

图 2.5 混合控制器结构

2.5.1 控制器设计

本节将设计由离散逆模型控制器和离散自适应滑模控制器组成的混合控制器。如图 2.6 所示,离散逆模型控制器包括 Hammerstein 系统线性环节的逆 $\hat{G}^{-1}$ 和非线性迟滞环节的逆 $\hat{f}^{-1}$ 两部分。上节讨论了 Hammerstein 系统的辨识,获得辨识结果后,$\hat{f}^{-1}$ 和 $\hat{G}^{-1}$ 容易求得。显然离散逆模型控制器是前馈补偿控制,该控制策略为开环控制,因此鲁棒性差,为了提高控制系统鲁棒性,设计了一种反馈控制器 - 离散自适应滑模控制器来解决此问题。

针对系统(1.1),若参考输入为 y_r,令 $\boldsymbol{R}(k)=\left[y_r(k), y_r(k-1), \cdots, y_r(k-n)\right]$,$\boldsymbol{Y}(k)=\left[y(k), y(k-1), \cdots, y(k-n)\right]$,则滑模面可定义如下:

$$s(k) = \boldsymbol{C}_e \boldsymbol{R}(k) - \boldsymbol{C}_e \boldsymbol{Y}(k) \qquad (2.23)$$

其中,$\boldsymbol{C}_e = \left[C^{n-1}, \cdots, C^1\right]$ 为滑模面的参数。

离散趋近律选择

$$s(k+1) = (1-qT)s(k) - \xi T \mathrm{sgn}(s(k)) \qquad (2.24)$$

其中,q 表示趋近滑模面的速度,即切换函数的动态过渡过程,$\xi>0$ 表示符号函数的增益,T 为采样周期。将式(2.24)改写为

$$s(k+1) = (1-qT)s(k) - \xi T\frac{s(k)}{|s(k)|} = \left(1-qT-\frac{\xi T}{|s(k)|}\right)s(k) = ps(k) \quad (2.25)$$

则有

$$p = 1 - qT - \frac{\xi T}{|s(k)|} \quad (2.26)$$

根据滑模面的选择与 Preisach 迟滞逆模型（2.17），离散自适应滑模控制器可以设计为

$$u(k) = f^{-1}\left((C_e B)^{-1}\left[C_e R(k+1) - C_e AY(k) - (1-qT)s(k) + \frac{q|s(k)|}{\eta}T\text{sgn}(s(k))\right]\right)$$

$$= \sum_{i=1}^{k}\delta_{ki}(C_e B)^{-1}\left[C_e R(i+1) - C_e AY(i) - (1-qT)s(i) + \frac{q|s(i)|}{\eta}T\text{sgn}(s(i))\right]$$

$$(2.27)$$

其中，η 为正常数，δ_{ki} 表示矩阵 δ 的 k 行 i 列元素，A，B 定义为

$$A = \begin{bmatrix} 0 & 1 & 0 & \cdots & 0 \\ 0 & 0 & 1 & \cdots & 0 \\ \cdots & \cdots & \cdots & & 1 \\ -1 & -a_1 & -a_2 & \cdots & -a_{n-1} \end{bmatrix}, B = \begin{bmatrix} b_1 \\ b_2 \\ \cdots \\ b_n \end{bmatrix} \text{①②} \quad (2.28)$$

---

① 离散自适应滑模控制器（2.27）中有三个重要参数 C_e，q 和 ξ 影响控制性能，C_e 决定了滑模的收敛速度，影响其动态响应，保证滑模运动渐近稳定且快速响应，C_e 越大，滑模运动段响应越快；q 决定着到达滑模面的方式，即切换函数的动态过程，可以改善系统运动品质；ξ 是符号函数增益，决定着滑模控制的克服摄动和外部干扰的能力，ξ 越大，抗干扰和摄动能力越强，但是其抖震幅度也越大。

② 如何减小滑模控制的抖震一直是滑模设计的重点问题，有不同方法可解决此问题，一些学者用双曲正切函数 tanh(.)、饱和函数 sat(.) 和其他连续函数代替符号函数 sign(.)。另一些学者则采用自适应方法来降低抖震。本章在式（2.27）中采用另一种自适应方法降低抖震的影响。设计自适应参数 $\frac{q|s(k)|}{\eta}$，则当滑模面 $s(k)$ 变化时可调节 ξ 将其变为正变量，此时抖震受自适应参数 $\frac{q|s(k)|}{\eta}$ 的影响减弱。图 2.6 给出了不同降低抖震方法比较的例子，从图中可以明确看出不同的降低抖震方法的效果。

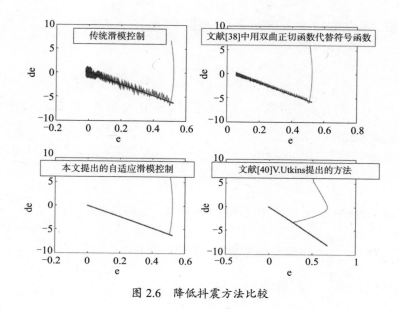

图 2.6 降低抖震方法比较

2.5.2 稳定性分析

根据图 2.5，取 $\varsigma > 0$ 为任意小的正实数，如果 $|u_r - u_s| > \varsigma$，表明离散自适应滑模控制器中滑模动态还未到达滑模面，此时选择离散逆模型控制，将加快到达滑模面时间；如果 $|u_r - u_s| \leq \varsigma$，说明滑模动态接近或已经到达滑模面，此时选择离散自适应滑模控制，将提高控制系统的鲁棒性。以下定理保证了离散自适应滑模控制的稳定性。

定理 2.2 对 Hammerstein 系统即式（1.1），若其线性环节如式（2.1），非线性环节由 Preisach 迟滞模型描述，滑模面选为式（2.23），趋近律为式（2.24），控制器选择为（2.27），（2.28），则 $|s(k)|$ 收敛到一个有界小邻域 Ω，$\Omega = \left\{ |s(k)| \Big| |s(k)| \leq \dfrac{\xi T}{2 - qT} \right\}$，且控制系统稳定。

证明 考虑式（2.26），有 $p<1$，根据（2.25），下式成立：

$$|p|=\frac{|s(k+1)|}{|s(k)|} \tag{2.29}$$

令 $|p|=1$，则有 $p=1$ 或者 $p=-1$。考虑到 $q>0$，$\xi>0$，只有 $p=-1$ 成立。因此，根据式（2.26），有：

$$|s(k)|=\frac{\xi T}{2-qT} \tag{2.30}$$

此时，$|s(k+1)|=|s(k)|$，滑动模态来回振荡，为临界状态。

当 $|s(k)|<\frac{\xi T}{2-qT}$，$|s(k)|$ 逐渐增大，系统不稳定。

当 $|s(k)|>\frac{\xi T}{2-qT}$，可得到下式：

$$\begin{aligned}|p|&<1\\|s(k+1)|&=|s(k)|\end{aligned} \tag{2.31}$$

$|s(k)|$ 逐渐减小，则系统稳定。

令 $\xi=\frac{q|s(k)|}{\eta}$，如果 $T<\frac{2n}{(n+1)q}$，趋近律可以改写为：

$$s(k+1)-s(k)=-qTs(k)-\frac{q|s(k)|}{\eta}T\operatorname{sign}(s(k)) \tag{2.32}$$

由式(2.32)可得：

$$\upsilon(k)=(\boldsymbol{C}_e\boldsymbol{B})^{-1}\left[\boldsymbol{C}_e\boldsymbol{R}(k+1)-\boldsymbol{C}_e\boldsymbol{A}\boldsymbol{Y}(k)-(1-qT)s(k)+\frac{q|s(k)|}{\eta}T\operatorname{sign}(s(k))\right] \tag{2.33}$$

考虑到 Preisach 迟滞非线性逆模型（2.17），可以得到控制器为：

$$u(k) = f^{-1}\left((C_e B)^{-1}\left[C_e R(k+1) - C_e A Y(k) - (1-qT)s(k) + \frac{q|s(k)|}{\eta}T\text{sign}(s(k)) \right] \right)$$

$$= \sum_{i=1}^{k} \delta_{ki}(C_e B)^{-1}\left[C_e R(i+1) - C_e A Y(i) - (1-qT)s(i) + \frac{q|s(i)|}{\eta}T\text{sign}(s(i)) \right]$$

（2.34）

为了验证控制系统的稳定性，选取李亚普诺夫函数为：

$$V(k) = |s(k)| \tag{2.35}$$

则

$$\Delta V(k) = |s(k+1)| - |s(k)| \tag{2.36}$$

当式（2.36）负定时，即 $|s(k+1)| < |s(k)|$ 时，控制系统稳定。

考虑式（2.32），如果 $s(k) > 0$，则 $s(k+1) < s(k)$；如果 $s(k) < 0$，则 $s(k+1) > s(k)$。因此，下式成立：

$$|s(k+1)| < |s(k)| \tag{2.37}$$

即控制系统稳定且 $|s(k)|$ 收敛到 Ω。

证毕。

2.6 仿真与实验

设计仿真验证所提出算法，仿真用迟滞数据是通过实验得到的离线数据，实验将通过含有迟滞的转台伺服系统验证。转台伺服系统（Googoltech）如图 2.7 所示，由驱动电机（HC-UFS13）、驱动卡（MR-J2S-10A，包括编码器和脉宽调制放大器）、数字信号处理器（TMS320 2812）和一台电脑（CPU：Pentium 2.8 GHz）组成。本章提出的算法均通过 C++ 和 CCS 3.0 软件实现。

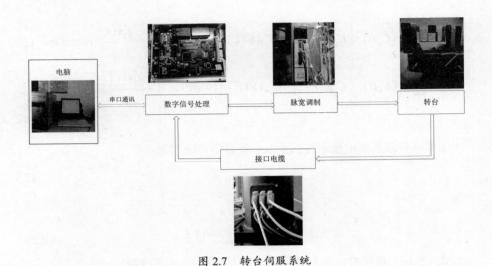

图 2.7 转台伺服系统

仿真用迟滞数据是通过实验得到的离线数据,其中采样间隔和输入电压分别为 $T=1s$,$U_{in}=5V$,传递函数为 $G(z)=\dfrac{(z+2)(z-5)}{(z+1)(z+3)(z+4)}$。本章提出的辨识 Hammerstein 系统线性环节的方法和文献[19]提出的方法比较,结果见表 2.1,其中 $h=0.25$,$\rho=4$。正如文献[19]所述,所提出方法更适合 Wiener-hammerstein 系统(误差最小),而不是 Hammerstein 系统(误差较小)。而本章所提出的方法更适合 Hammerstein 系统,从表 2.1 可以明确看出本书所提出方法与文献[19]相比,误差更小。

表 2.1 两种辨识方法结果比较

	a_1	a_2	a_3	b_1	b_2
本书方法	7.9965	19.0401	12.0012	2.9916	10.0111
文献[19]方法	8.1102	19.1044	11.9817	2.9863	10.0225

为了验证本章所提出控制策略的有效性，比较了离散逆模型控制、离散自适应滑模控制、自适应控制和混合控制，如图 2.8 至图 2.11 所示。由于采用仿真验证，对传递函数 $G(z)=\dfrac{(z+2)(z-5)}{(z+1)(z+3)(z+4)}$ 求逆，并根据式（2.16）和式（2.23）在输入信号为 $y=\sin(2\pi t)$ 时计算 Preisach 逆模型并求出 δ。

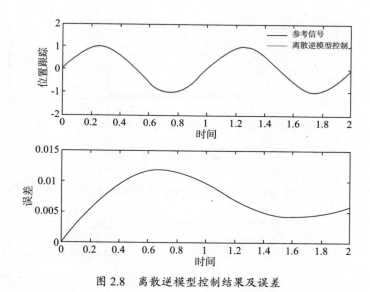

图 2.8　离散逆模型控制结果及误差

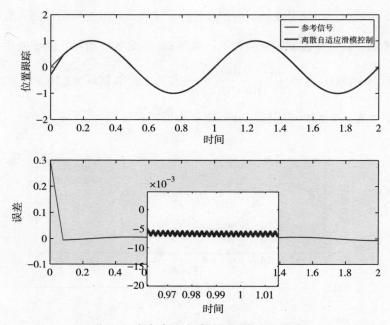

图 2.9 离散自适应滑模控制结果及误差

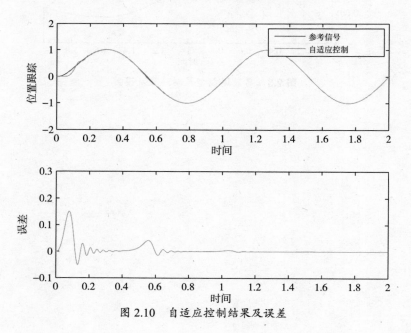

图 2.10 自适应控制结果及误差

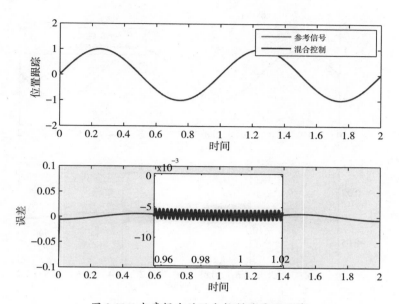

图 2.11 本章提出的混合控制结果及误差

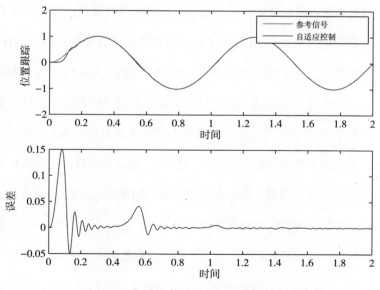

图 2.12 高斯噪声自适应控制结果及误差

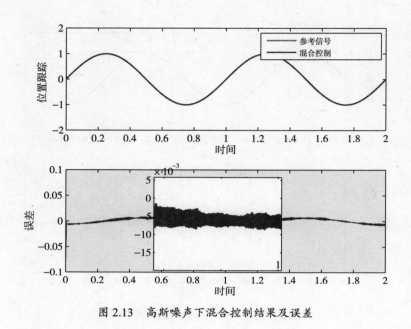

图 2.13　高斯噪声下混合控制结果及误差

由图 2.9 至图 2.11 可以看出，自适应控制比离散自适应滑模控制有更好的瞬态响应（或跟踪误差更小），但是比不上本章提出的混合控制策略。所有控制中，混合控制策略指标最优（如图 2.11 所示，平均绝对误差（MAE）为 0.0042）。

另外，自适应控制和离散自适应滑模控制有相似的稳态响应，其稳态性能相差不大（如图 2.10 所示，自适应控制的平均绝对误差为 0.0076，而离散自适应滑模控制平均绝对误差为 0.0087，如图 2.9 所示），但是离散自适应滑模控制收敛时间比自适应控制更快（离散自适应滑模控制收敛时间大约 0.1 s，而自适应控制收敛时间大约为 0.2 s，见图 2.9 和图 2.10。

图 2.12 和图 2.13 分别是高斯噪声下自适应控制和混合控制的跟踪曲线与误差。从图中可以看出自适应控制和混合控制均具有较强的鲁棒性，但是显然混合

控制不论是暂态性能还是稳态性能及控制精度均好于自适应控制（高斯噪声下自适应控制的平均绝对误差为 0.0078，而混合控制的为 0.0045），更适合实际应用。

2.7　结论

本章针对迟滞 Hammerstein 系统采用离散 Preisach 模型描述其迟滞非线性，并放松了盲辨识法辨识 Hammerstein 系统的条件，在系统阶次未知的情况下，用 Hankel 矩阵首先辨识系统阶次；然后在只有系统输入输出数据已知而无其他已知参数情况下，采用过采样盲辨识法辨识 Hammerstein 系统线性环节；根据辨识结果求得线性环节输入即非线性环节输出数据，在此基础上应用下三角矩阵法辨识离散 Preisach 模型从而估计 Hammerstein 系统。新提出的下三角矩阵辨识法仅需要 m 维 Preisach 算子小于其他辨识算法，因而提高了计算效率。模型辨识以后为了更加精确地控制 Hammerstein 系统，提出了由离散逆模型控制和离散自适应滑模控制组成的混合控制策略并保证了控制系统的稳定性。该控制策略不仅可以迅速到达稳态并且具有很好的鲁棒性，仿真结果证明了所用方法的可行性和有效性。

参考文献

［1］MAYERGOYZ ID. Mathematical models of hysteresis［M］. New York: Springer-verlag.1991.

［2］TAN XB, JOHN SB. Adaptive Identification and Control of Hysteresis in Smart Materials［J］. IEEE Transactions on Automatic Control, 2005,50(6):827-839.

[3] GUO YX, MAO JQ, ZHOU KM. Rate-dependent mode-ling and control of GMA based on Hammerstein model with Preisach operator [C]. Mechatronics and Automation (ICMA), 2012 International Conference on, 2012:343-347.

[4] BAO KN, KYOUNG KA. Feedforward Control of Shape Memory Alloy Actuators Using Fuzzy-Based Inverse Preisach Model [J]. IEEE Transactions on Control Systems Technology, 2009,17(2):434-441.

[5] LI CT, TAN YH. Adaptive output feedback control of systems preceded by the Preisach-type hysteresis [J]. IEEE Transactions on Systems, Man, and Cybernetics, Part B: Cybernetics, 2005,35(1):130-135.

[6] MA YH, MAO JQ, ZHANG Z. On Generalized Dynamic Preisach Operator with Application to Hysteresis Nonlinear Systems [J]. IEEE Transactions on Control Systems Technology, 2011,19(6):1527-1533.

[7] TRAPANESE M. Identification of the Parameters of Reduced Vector Preisach Model by Neural Networks [J]. IEEE Transactions on Magnetics, 2008,44(11):3197-3200.

[8] BAO KN, KYOUNG KA. Feedforward Control of Shape Memory Alloy Actuators Using Fuzzy-Based Inverse Preisach Model [J]. IEEE Transactions on Control Systems Technology,2009,17(2):434-441.

[9] BAI EW, FU MY. A blind approach to Hammerstein model identification [J]. IEEE Transactions on Signal Processing, 2002,50(7):1610-1619.

[10] YU CP, ZHANG CS, XIE LH. A New Deterministic Identification Approach

to Hammerstein Systems [J]. IEEE Transactions on Signal Processing, 2014,62(1):131-140.

[11] LI GQ, WEN CY. Convergence of fixed-point iteration for the identification of Hammerstein and Wiener systems [J]. International Journal of Robust and Nonlinear Control, 2013,23(13):1510-1523.

[12] CERONE V, PIGA D, REGRUTO D. Computational Load Reduction in Bounded Error Identification of Hammerstein Systems [J]. IEEE Transactions on Automatic Control, 2013,58(5):1317-1322.

[13] DHAIFLLAH M, WESTWICK DT. Identification of Auto-Regressive Exogenous Hammerstein Models Based on Support Vector Machine Regression [J]. IEEE Transactions on Control Systems Technology, 2013,21(6):2083-2090.

[14] REN XM, LV XH. Identification of Extended Hammerstein Systems Using Dynamic Self-Optimizing Neural Networks [J]. IEEE Transactions on Neural Networks, 2011,22(8):1169-1179.

[15] MICHALKIEWICZ J. Modified Kolmogorov's Neural Network in the Identification of Hammerstein and Wiener Systems [J]. IEEE Transactions on Neural Networks and Learning Systems, 2012,23(4):657-662.

[16] BAI EW, LI QY, SOURA D. Blind identifiability of IIR systems [J]. Automatica, 2002,38(1):181-184.

[17] WANG JD, AKIRA S, DAVID S, et al. A blind approach to closed-loop identification of Hammerstein systems [J]. International Journal of Control,

2007,80(2):302-313.

[18] MOHAMMAD AJ, ANTHONY DA, DENNIS B. Retrospective-Cost Adaptive Control of Uncertain Hammerstein-Wiener Systems with Memoryless and Hysteretic Nonlinearities [C]. AIAA Guidance, Navigation, and Control Conference, 2012:2012-4449.

[19] TAN AH, WONG H, GODFREY KR. Identification of a Wiener-Hammerstein system using an incremental nonlinear optimisation technique [J]. Control Engineering Practice, 2012,20(11):1140-1148.

[20] 方崇智，萧德云. 过程辨识 [M]. 北京：清华大学出版社，2007.

[21] BAI EW, FU MY. Blind system identification and channel equalization of IIR systems without statistical information [J]. IEEE Transactions on Signal Processing, 1999,47(7):1910-1921.

第 3 章　Backlash-like 迟滞非线性系统预设精度自适应控制

3.1　引言

本章将对用 Backlash 类迟滞模型描述的 Hammerstein 系统进行研究，对于 Backlash 类迟滞模型，许多学者做了深入研究。王焕钦等学者[1]研究了非严格反馈的随机非线性系统，提出了自适应神经网络控制采用反步法解决含有 Backlash 类迟滞的非严格反馈的随机非线性系统。而刘烨和林岩[2]则应用高增益观测器和反步法处理 Backlash 类迟滞描述的非线性系统，并在不需要逆迟滞模型的条件下设计稳定的模糊控制器控制该系统。普利肯·马蒂厄等学者[3]研究了输入为 Backlash 或切换非线性的 Hammerstein 系统的子空间辨识问题，设计特殊输入信号估计系统非线性环节以完成子空间辨识。董瑞丽和谭永红[4]研究了离散 Hammerstein 系统的内模控制问题，其非线性环节采用离散 Backlash 模型描述，该内模控制基于逆 Hammerstein 模型补偿并设计滤波器保证控制系统的鲁棒性。但是几乎所有文献均未涉及用 Backlash 类模型描述的迟滞 Hammerstein 系统的规定性能控制问题。

规定性能控制最早由贝奇洛利斯和罗维塔基斯[5]提出,用来保证多输入多输出反馈线性化非线性系统的规定性能控制。那靖等学者[6]应用规定性能函数对严格反馈的非线性动态系统采用反步法和自适应模糊控制,定义新的虚拟变量保证控制误差收敛于规定性能函数范围内。而赵新龙等学者[7]将规定性能函数应用于Bouc-Wen迟滞非线性系统,通过Fourier变换取得近似解,然后提出规定性能自适应控制保证该系统的跟踪误差的暂态和稳态性能。那靖等学者[6]在文献[5]研究的基础上进一步改进规定性能函数,提出了一种新的规定性能函数形式,并将其应用于带摩擦的伺服电机系统设计自适应控制保证系统跟踪误差的稳态和暂态性能收敛于该规定性能函数内。

本章在文献[6]研究的基础上,进一步简化规定性能函数,提出一种新的函数并将其首次应用于Backlash类迟滞Hammerstein系统的自适应控制。为了简化控制器设计,首先将矢量误差转化为标量误差,然后将该误差通过新提出的简化的规定性能函数转化为规定性能误差。由于误差转化降低了控制性能,因此使用本章简化的规定性能函数约束误差转化后的控制精度,设计模型参考自适应控制器并通过Lyapunov函数保证闭环系统的稳定性。在新的规定性能函数的保证下,通过误差转换简化了控制器设计,又避免了误差转换带来的性能降低,并使暂态和稳态跟踪误差收敛于规定性能范围内。

主要创新总结如下:

(1)进一步简化文献[6]的规定性能函数,提出新的更加实用的规定性能函数。

（2）将矢量误差首先转换为标量误差，由于误差转换会降低系统性能，因此应用新提出的规定性能函数进一步对误差进行转换，保证所设计的自适应控制器跟踪误差收敛于规定性能函数范围内。做到既简化控制器设计，又保证系统控制精度。

（3）为了分析闭环系统的稳定性，设计 Lyapunov 函数并通过 Lambert-W 函数证明该系统稳定且暂态和稳态跟踪误差收敛于规定范围。

3.2 问题描述

仍然考虑 Hammerstein 系统，用 Backlash 类模型描述迟滞非线性，表达式如下：

$$\begin{cases} \dot{X}(t) = AX(t) + Bu(t) \\ \dot{\upsilon}(t) = \alpha|\dot{u}(t)|(cu(t) - \upsilon(t)) + d\dot{u}(t) \end{cases} \quad (3.1)$$

其中，$A \in \mathbb{R}^{n \times n}$，$B \in \mathbb{R}^{n \times 1}$ 是未知系统参数；$X(t) \in \mathbb{R}^{n \times 1}$，$\upsilon(t) \in \mathbb{R}^1$ 为系统状态和系统线性环节输入；c，d 是常数，有 $c>0$ 为迟滞斜率且满足 $c>d$；$\acute{a}$ 是正系数；$u(t)$ 是系统的输入信号。

Backlash 类迟滞用 $\dot{\upsilon}(t) = \alpha|\dot{u}(t)|(cu(t) - \upsilon(t)) + d\dot{u}(t)$ 描述 Hammerstein 系统非线性环节，其输出 $v(t)$ 同时是线性环节输入。图 3.1 所示的是当 α=1.2，c=4.5，d=0.65，且输入为 $u(t)=K\sin(3.5t)$，K=2.5 和 K=4.5 时 Backlash 类迟滞的曲线图。

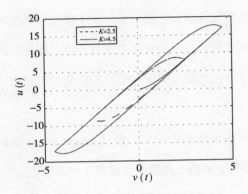

图 3.1　Backlash 类迟滞曲线

根据文献[8]，可求得式 $\dot{v}(t)=\alpha|\dot{u}(t)|(cu(t)-v(t))+d\dot{u}(t)$ 的解析解：

$$v(t)=cu(t)+h[u(t)] \tag{3.2}$$

其中，

$$h[u(t)]=(\tilde{v}_0-cu_0)e^{-\alpha(u(t)-u_0)\operatorname{sign}(\dot{u}(t))}+e^{-\alpha u(t)\operatorname{sign}(\dot{u}(t))}\int_{v_0}^{u(t)}(d-c)e^{\alpha\xi\operatorname{sign}(\dot{u}(t))}\mathrm{d}\xi \tag{3.3}$$

将式（3.2）带入系统（3.1），得：

$$\dot{X}(t)=AX(t)+Bcu(t)+Bh[u(t)] \tag{3.4}$$

为了对式（3.4）设计模型参考自适应控制器，定义参考模型为：

$$\dot{X}_r(t)=A_rX_r(t)+B_rv_r(t) \tag{3.5}$$

其中，$u_r(t)$ 是参考输入，$\boldsymbol{A}_r\in\mathbb{R}^{n\times n}$ 和 $\boldsymbol{B}_r\in\mathbb{R}^{n\times 1}$。

本章目标为：

（1）根据参考模型（3.5），将矢量误差转化为标量误差；

（2）进一步简化规定性能函数，提出一种新的更加简单的规定性能函数；

（3）在新规定性能函数下，设计模型参考自适应控制器，并应用 Lambert-W 函数保证闭环系统的稳定控制且稳态和暂态误差收敛于规定性能范围内。

3.3 控制器设计

在本节中,将设计模型参考自适应控制器控制 Backlash 类迟滞 Hammerstein 系统,所提出的自适应控制器可保证闭环系统稳定且跟踪误差收敛于规定性能范围内。该控制器包括两部分:自适应控制信号 $u_a(t)$ 和参考输入信号 $u_r(t)$。

3.3.1 误差转换

为了设计自适应控制器,需要将矢量误差转换为标量误差。假设存在常向量 $\boldsymbol{\theta}_x^* \in \mathbb{R}^{n\times 1}$ 和非零常标量 θ_r^* 满足下面方程:

$$\boldsymbol{A} + \boldsymbol{B}\boldsymbol{\theta}_x^{*T} = \boldsymbol{A}_r, \quad \boldsymbol{B}\theta_r^* = \boldsymbol{B}_r, \tag{3.6}$$

其中, θ_r^* 的符号是已知的。不失一般性,可假设 θ_r^* 为正。则根据文献[9],有如下引理:

引理 3.1 令

$$\begin{cases} \dot{\boldsymbol{X}}(t) = \boldsymbol{A}\boldsymbol{X}(t) + \boldsymbol{b}\boldsymbol{m}(t) \\ \lambda(s) = (s+k)\boldsymbol{R}(s) \end{cases} \tag{3.7}$$

其中, $\boldsymbol{A}, \boldsymbol{b}$ 是可控的,系统渐进稳定,其特征多项式为 $\lambda(s)$,且 $k>0$,则有:

(1) 存在 w 满足

$$w^T(s\boldsymbol{I} - \boldsymbol{A})^{-1}b = \frac{1}{s+k} \tag{3.8}$$

（2）如果 $x=w^T X$，那么（Ⅰ）：$x \in L^\infty \Rightarrow X \in L^\infty$，（Ⅱ）：如果 $\lim_{t\to\infty} x(t)=0$，则 $\lim_{t\to\infty} X(t)=0$.①

如前所述，本章将模型参考控制输入分为两部分：自适应控制信号 $u_a(t)$ 和参考输入信号 $u_r(t)$，因此定义 $u(t)$ 如下：

$$u(t)=u_a(t)+u_r(t), \tag{3.9}$$

同时定义误差为 $E(t) = X(t) - X_r(t)$。根据（3.2）（3.3）（3.4）（3.5）（3.6）和（3.9），E 的时间导数可以表示为

$$\begin{aligned}\dot{E}(t) &= \dot{X}(t) - \dot{X}_r(t) = Ax(t) + Bcu(t) + Bh[u(t)] - A_r X_r(t) - B_r u_r(t) \\ &= A_r E(t) + B\left[cu_a(t) + (c-\theta_r^*)u_r(t) - \theta_x^{*T} x(t) + h[u(t)]\right].\end{aligned} \tag{3.10}$$

根据引理 3.1，显然从（3.8）中可表明存在 w，使得

$$w^T(sI-A_r)^{-1}B = \frac{1}{s+k} \tag{3.11}$$

则状态矢量误差 $E(t)$ 可以转换为如下标量误差：

$$e_s(t) = w^T E(t). \tag{3.12}$$

对式（3.10）进行 Laplace 变换，复频域下矢量误差可以表示为

$$E(s) = (sI-A_r)^{-1} B\left[cu_a(s) + (c-\theta_r^*)u_r(s) - \theta_x^{*T} x(s) + h[u(s)]\right] \tag{3.13}$$

---

① 引理 3.1 可以十分方便地将一个可控且渐进稳定系统的矢量误差转化为标量误差，为了简化控制器设计，我们应用引理 3.1 将 Backlash 类迟滞 Hammestein 系统（3.4）的状态矢量误差转化为标量误差。从后面仿真（图 3.3 和图 3.4；图 3.7 和图 3.8）可以看出，采用引理 3.1 将矢量误差转换为标量后，标量误差与状态 $x_1(t)$ 的误差 $e(t)$ 类似，但是绝对误差较 $e_1(t)$ 大，因此从某种意义上说，矢量误差转化为标量误差虽然简化了设计，但是却降低了精度（对 $y(t)=x_1(t)$ 来说，该结论显然成立，实际纯反馈系统均可以表示成 $y(t)=x_1(t)$ 形式）。

等式两边同乘以 $\boldsymbol{w}^{\mathrm{T}}$，得：

$$\boldsymbol{w}^{\mathrm{T}}\boldsymbol{E}(s) = \boldsymbol{w}^{\mathrm{T}}(s\boldsymbol{I}-\boldsymbol{A}_r)^{-1}\boldsymbol{B}\left[cu_a(s)+(c-\theta_r^*)u_r(s)-\boldsymbol{\theta}_x^{*\mathrm{T}}\boldsymbol{x}(s)+h[u(s)]\right]$$
（3.14）

根据式（3.11）和式（3.12），可以将式（3.14）写为：

$$e_s(s) = \frac{1}{(s+k)}\left[cu_a(s)+(c-\theta_r^*)u_r(s)-\boldsymbol{\theta}_x^{*\mathrm{T}}\boldsymbol{x}(s)+h[u(s)]\right]$$
（3.15）

对式（3.15）进行 Laplace 反变换，得到时域下标量误差为：

$$\dot{e}_s(t) = -ke_s(t)+\left[cu_a(t)+(c-\theta_r^*)u_r(t)-\boldsymbol{\theta}_x^{*\mathrm{T}}\boldsymbol{x}(t)+h[u(t)]\right]$$
（3.16）

3.3.2 规定性能函数

为了使误差 $e_s(t)$ 收敛于规定性能范围内，需首先定义规定性能函数。定义一个正光滑递减函数 $\rho(t)$：$\mathbb{R}^+ \to \mathbb{R}^+$，且有 $\lim_{t\to\infty}\rho(t)=\rho_\infty>0$ 作为规定性能函数，该函数保证如下约束条件成立：

$$-\underline{\delta}\rho(t) < e_s(t) < \overline{\delta}\rho(t)$$
（3.17）

其中，规定尺度参数 $\underline{\delta}$ 和 $\overline{\delta}$ 是设计人员选择的正常数。

如文献[10, 11, 12-14]所述，我们选择 $\rho(t)$ 如下：

$$\rho(t) = (\rho_0-\rho_\infty)\mathrm{e}^{-\beta t}+\rho_\infty$$
（3.18）

其中，$\rho_0>\rho_\infty$ 和 $\beta>0$ 是合适的常数。

根据文献[11, 13-17]的研究，定义规定转换误差 $z\in\mathbb{R}$ 的严格单增函数 $S(z)$ 为：$-\underline{\delta}<S(z)<\overline{\delta}$，$\lim_{z\to+\infty}S(z)=\overline{\delta}$，$\lim_{z\to-\infty}S(z)=-\underline{\delta}$ 则条件（3.17）可以表示为：

$$e_s(t) = \rho(t)S(z)$$
（3.19）

在文献[14]研究的基础上，我们重新定义 $S(z)$ 为

$$S(z) = \frac{\overline{\delta}e^z - \underline{\delta}}{e^z + 1}. \tag{3.20}$$

式（3.20）为新定义的简化的规定性能函数。又由于 $S(z)$ 是光滑、严格单增的函数，其反函数存在且可求得：

$$z = S^{-1}\left[\frac{e_s(t)}{\rho(t)}\right] = \ln\left(\frac{\frac{e_s(t)}{\rho(t)} + \underline{\delta}}{\overline{\delta} - \frac{e_s(t)}{\rho(t)}}\right). \tag{3.21}$$

根据式（3.21）可进一步推导出规定转换误差 z 的导数如下：

$$\dot{z} = \frac{(\overline{\delta} + \underline{\delta})(\dot{e}_s(t)\rho(t) - e_s(t)\dot{\rho}(t))}{(\underline{\delta}\rho(t) + e_s(t))(\overline{\delta}\rho(t) - e_s(t))} \tag{3.22}$$

考虑到（3.18），有

$$\dot{\rho}(t) = -\beta\rho(t) + \beta\rho_\infty \tag{3.23}$$

又应用式（3.16）和式（3.23），式（3.22）可以推导为

$$\dot{z} = \varLambda\left(\gamma(t)e_s(t) + c\rho(t)\upsilon_a(t) + (c - \theta_r^*)\rho(t)\upsilon_r(t) - \rho(t)\boldsymbol{\theta}_x^{*\mathrm{T}}\boldsymbol{X}(t) + \rho(t)h[\upsilon(t)]\right), \tag{3.24}$$

其中，$\varLambda = \dfrac{\overline{\delta} + \underline{\delta}}{(\underline{\delta}\rho(t) + e_s(t))(\overline{\delta}\rho(t) - e_s(t))}$ 且 $\gamma(t) = \beta\rho(t) - k\rho(t) - \beta\rho_\infty$。

根据式（3.17）和（3.18）可知 $\varLambda > 0$，又根据式（3.3）（3.9），可以将式（3.24）改写为

$$\dot{z} = \Lambda \begin{pmatrix} \gamma(t)e_s(t) + c\rho(t)v_a(t) + (c-\theta_r^*)\rho(t)v_r(t) - \rho(t)\theta_x^{*T}X(t) \\ +\rho(t)\left((u_0 - cv_0)e^{-\alpha(v(t)-v_0)\text{sign}(\dot{v}(t))} + e^{-\alpha v(t)\text{sign}(\dot{v}(t))}\int_{v_0}^{v(t)}(d-c)e^{\alpha\xi\text{sign}(\dot{v}(t))}d\xi \right) \end{pmatrix}$$

$$= \Lambda \begin{pmatrix} \gamma(t)e_s(t) + c\rho(t)v_a(t) + (c-\theta_r^*)\rho(t)v_r(t) - \rho(t)\theta_x^{*T}X(t) \\ +\rho(t)\left((u_0 - cv_0)e^{-\alpha(v_a(t)+v_r(t)-v_0)\text{sign}(\dot{v}(t))} + e^{-\alpha(v_a(t)+v_r(t))\text{sign}(\dot{v}(t))}\int_{v_0}^{v(t)}(d-c)e^{\alpha\xi\text{sign}(\dot{v}(t))}d\xi \right) \end{pmatrix}$$

（3.25）

3.3.3 控制器设计

为了精确控制该 Hammerstein 系统，定义变量如下：

$$\begin{cases} \tilde{\gamma}(t) = \gamma(t) - \hat{\gamma}(t) \\ \tilde{\theta}_r = \theta_r^* - \hat{\theta}_r \\ \tilde{\theta}_x = \theta_x^* - \hat{\theta}_x \end{cases} \quad （3.26）$$

其中，$\hat{\gamma}(t)$，$\hat{\theta}_r$，$\hat{\theta}_x$ 分别是 $\gamma(t)$ $\gamma(t)$，θ_r^* 和 θ_x^* 的估计值。

对于本章给定的 Backlash 类迟滞 Hammerstein 系统，可以将自适应控制器设计为

$$u_a(t) = \frac{(\hat{\theta}_r - c)\rho(t)u_r(t) + \rho(t)\hat{\theta}_x^T x(t) - \hat{\gamma}(t)e_s(t) - \phi z}{c\rho(t)} + \frac{W\left(\dfrac{-T\alpha\text{sign}\dot{u}(t)}{c\rho(t)} e^{\frac{R\alpha\text{sign}\dot{u}(t)}{c\rho(t)}} \right)}{\alpha\text{sign}(\dot{u}(t))},$$

（3.27）

其中，

$$T = \rho(t)\left((v_0 - cu_0)e^{-\alpha(u_r(t)-u_0)\text{sign}\dot{u}(t)} + e^{-\alpha u_r(t)\text{sign}\dot{u}(t)}\int_{u_0}^{u(t)}(d-c)e^{\alpha\xi\text{sign}\dot{u}(t)}d\xi \right)$$

$$R = \hat{\gamma}(t)e_s(t) + (c-\hat{\theta}_r)\rho(t)u_r(t) - \rho(t)\hat{\theta}_x^T x(t) - \phi z \quad （3.28）$$

此处 $\phi > 0$ 且 $W(.)$ 表示 Lambert-W 函数。

自适应控制率设计为

$$\begin{cases} \dot{\hat{\gamma}} = \Lambda z e_s(t) + (\beta k - \beta^2)\rho(t) + (\beta k - \beta^2)\rho_\infty \\ \dot{\hat{\theta}}_r = \Lambda z \rho(t) \upsilon_r(t) \\ \dot{\hat{\theta}}_x = \Lambda z \rho(t) x(t) \end{cases} \quad (3.29)$$

且

$$T\alpha\text{sign}\dot{u}(t) \geqslant 0, e^{\left(2\frac{R\alpha\text{sign}\dot{u}(t)}{c\rho(t)}\right)} \geqslant c\rho(t). \quad (3.30)$$

设计自适应控制器和控制率后，如下定理保证所设计控器使得闭环系统稳定。

定理 3.1 对含有迟滞（3.2）和（3.3）的系统（3.1），如果自适应控制器和自适应控制率分别设计为（3.27）（3.28）（3.29）和（3.30），则闭环控制系统稳定。

证明 选取 Lyapunov 函数如下：

$$V(t) = \frac{1}{2}z^2 + \tilde{\boldsymbol{\theta}}_x^\mathrm{T}\tilde{\boldsymbol{\theta}}_x + \frac{1}{2}\tilde{\gamma}^2(t) + \frac{1}{2}\tilde{\theta}_r^2 \quad (3.31)$$

根据式（3.24），可以求得 $V(t)$ 的导数为：

$$\begin{aligned} \dot{V}(t) = &z\Lambda\big(\gamma(t)e_s(t) + c\rho(t)u_a(t) + (c-\theta_r^*)\rho(t)u_r(t) - \rho(t)\boldsymbol{\theta}_x^{*\mathrm{T}}\boldsymbol{x}(t) \\ &+ \rho(t)h[u(t)] + \tilde{\gamma}(t)\dot{\tilde{\gamma}}(t) + \tilde{\theta}_r \dot{\tilde{\theta}}_r + \tilde{\boldsymbol{\theta}}_x^\mathrm{T}\dot{\tilde{\boldsymbol{\theta}}}_x\big) \end{aligned} \quad (3.32)$$

将式（3.3）（3.26）和（3.27）代入式（3.32）可得：

$$\begin{aligned}
\dot V(t) &= \tilde\gamma(t)\Big(\Lambda z e_s(t)+\tilde\gamma(t)\Big)+\tilde\theta_r\Big(\Lambda z\rho(t)u_r(t)+\dot{\tilde\theta}_r\Big)+\tilde\theta_x^{\mathrm T}\Big(\Lambda z\rho(t)\boldsymbol x(t)+\dot{\tilde\theta}_x\Big)\\
&\quad -\Lambda\phi z^2+\dfrac{W\left(\dfrac{-T\alpha\operatorname{sign}\dot u(t)}{c\rho(t)}\mathrm e^{\frac{R\alpha\operatorname{sign}\dot u(t)}{c\rho(t)}}\right)}{\alpha\operatorname{sign}\dot u(t)}+\rho(t)h[\upsilon(t)]\\
&= -\Lambda\phi z^2+\tilde\gamma(t)\Big(\Lambda z e_s(t)+(\beta k-\beta^2)\rho(t)+(\beta^2-\beta k)\beta_\infty-\dot{\hat\gamma}(t)\Big)\\
&\quad +\tilde\theta_r\Big(\Lambda z\rho(t)u_r(t)-\dot{\hat\theta}_r\Big)+\tilde\theta_x^{\mathrm T}\Big(\Lambda z\rho(t)\boldsymbol x(t)-\dot{\hat\theta}_x\Big)\\
&\quad +\dfrac{W\left(\dfrac{-T\alpha\operatorname{sign}\dot u(t)}{c\rho(t)}\mathrm e^{\frac{R\alpha\operatorname{sign}\dot u(t)}{c\rho(t)}}\right)}{\alpha\operatorname{sign}\dot u(t)}+T\mathrm e^{-\alpha\operatorname{sign}\dot u(t)u_a(t)}
\end{aligned} \quad (3.33)$$

又根据式（3.29），有：

$$\begin{aligned}
\dot V &= -\Lambda\phi z^2+\dfrac{W\left(\dfrac{-T\alpha\operatorname{sign}\dot u(t)}{c\rho(t)}\mathrm e^{\frac{R\alpha\operatorname{sign}\dot u(t)}{c\rho(t)}}\right)}{\alpha\operatorname{sign}\dot u(t)}+T\mathrm e^{-\alpha\operatorname{sign}\dot u(t)u_a(t)}\\
&= -\Lambda\phi z^2+\dfrac{W\left(\dfrac{-T\alpha\operatorname{sign}\dot u(t)}{c\rho(t)}\mathrm e^{\frac{R\alpha\operatorname{sign}\dot u(t)}{c\rho(t)}}\right)}{\alpha\operatorname{sign}\dot u(t)}+T\mathrm e^{-\alpha\operatorname{sign}\dot u(t)\left(\dfrac{R}{c\rho(t)}+\dfrac{W\left(\dfrac{-T\alpha\operatorname{sign}\dot u(t)}{c\rho(t)}\mathrm e^{\frac{R\alpha\operatorname{sign}\dot u(t)}{c\rho(t)}}\right)}{\alpha\operatorname{sign}\dot u(t)}\right)}
\end{aligned}\quad (3.34)$$

定义如下等式：

$$S=\alpha\operatorname{sign}\dot u(t),\ J=c\rho(t) \quad (3.35)$$

则

$$\dfrac{W\left(\dfrac{-T\alpha\operatorname{sign}\dot u(t)}{c\rho(t)}\mathrm e^{\frac{R\alpha\operatorname{sign}\dot u(t)}{c\rho(t)}}\right)}{\alpha\operatorname{sign}\dot u(t)}+T\mathrm e^{-\alpha\operatorname{sign}\dot u(t)\left(\dfrac{R}{c\rho(t)}+\dfrac{W\left(\dfrac{-T\alpha\operatorname{sign}\dot u(t)}{c\rho(t)}\mathrm e^{\frac{R\alpha\operatorname{sign}\dot u(t)}{c\rho(t)}}\right)}{\alpha\operatorname{sign}\dot u(t)}\right)}=0. \quad (3.36)$$

将（3.35）代入（3.36），我们有

$$\frac{W\left(\dfrac{-TS}{J}e^{\frac{RS}{J}}\right)}{S} = -Te^{-S\left(\frac{R}{J}+\frac{W\left(\frac{-TS}{J}e^{\frac{RS}{J}}\right)}{S}\right)} \qquad (3.37)$$

等式两边同乘以 $e^{W\left(\frac{-TS}{J}e^{\frac{RS}{J}}\right)}$，（3.37）可改写为

$$W\left(\frac{-TS}{J}e^{\frac{RS}{J}}\right)e^{W\left(\frac{-TS}{J}e^{\frac{RS}{J}}\right)} = -TSe^{-\frac{RS}{J}} \qquad (3.38)$$

因此，可以得到

$$W\left(\frac{-TS}{J}e^{\frac{RS}{J}}\right) = W\left(-TSe^{-\frac{RS}{J}}\right) \qquad (3.39)$$

考虑到式（3.30），有根据 Lambert-W 函数的性质，下面不等式成立：

$$W\left(\frac{-TS}{J}e^{\frac{RS}{J}}\right) \leqslant W\left(-TSe^{-\frac{RS}{J}}\right) \qquad (3.40)$$

因此有

$$\frac{W\left(\dfrac{-T\alpha\operatorname{sign}\dot{u}(t)}{c\rho(t)}e^{\frac{R\alpha\operatorname{sign}\dot{u}(t)}{c\rho(t)}}\right)}{\alpha\operatorname{sign}\dot{u}(t)} + Te^{-\alpha\operatorname{sign}\dot{u}(t)\left(\frac{R}{c\rho(t)}+\frac{W\left(\frac{-T\alpha\operatorname{sign}\dot{u}(t)}{c\rho(t)}e^{\frac{R\alpha\operatorname{sign}\dot{u}(t)}{c\rho(t)}}\right)}{\alpha\operatorname{sign}\dot{u}(t)}\right)} \leqslant 0. \qquad (3.41)$$

令

$$\frac{W\left(\dfrac{-T\alpha\operatorname{sign}\dot{u}(t)}{c\rho(t)}e^{\frac{R\alpha\operatorname{sign}\dot{u}(t)}{c\rho(t)}}\right)}{\alpha\operatorname{sign}\dot{u}(t)} + Te^{-\alpha\operatorname{sign}\dot{u}(t)\left(\frac{R}{c\rho(t)}+\frac{W\left(\frac{-T\alpha\operatorname{sign}\dot{u}(t)}{c\rho(t)}e^{\frac{R\alpha\operatorname{sign}\dot{u}(t)}{c\rho(t)}}\right)}{\alpha\operatorname{sign}\dot{u}(t)}\right)} = -x^2 \qquad (3.42)$$

则
$$\dot{V} = -\Lambda\phi z^2 - x^2 \tag{3.43}$$

由式（3.27），式（3.31）和（3.43）可知 V 非增，因此闭环系统稳定。又由于（3.17）和（3.18）的约束，则标量误差 $e_s(t)$ 收敛于规定性能范围内。

得证。

3.4 仿真

为了验证本章所提出新的规定性能函数和自适应控制器的有效性和正确性，考虑系统如 (3.1)，取值如下：

$$\boldsymbol{A} = \begin{bmatrix} 0 & 1 \\ -6 & -7 \end{bmatrix}, \boldsymbol{B} = \begin{bmatrix} 0 \\ 8 \end{bmatrix}$$

为了设计模型参考自适应控制器，参考模型 (3.5) 系数取值如下

$$\boldsymbol{A}_r = \begin{bmatrix} 0 & 1 \\ -10 & -5 \end{bmatrix}, \boldsymbol{B}_r = \begin{bmatrix} 0 \\ 2 \end{bmatrix}$$

根据以上信息可以计算出转换误差的各个参数为：$\boldsymbol{\theta}_x^* = \begin{bmatrix} -\frac{1}{2} & \frac{1}{4} \end{bmatrix}^{\mathrm{T}}$，$\theta_r^* = \frac{1}{4}$，$\boldsymbol{w} = \begin{bmatrix} \frac{1}{48} & \frac{1}{8} \end{bmatrix}^{\mathrm{T}}$，$k=1$。分别选取参考输入 $u_r(t) = 2\sin(2.5\pi t)$ 和 $u_r(t) = 2\sin(2\pi t)$，Backlash 类迟滞模型和规定性能函数的参数取值如表 3.1 所示。控制结果如图 3.2～3.9 所示。

表 3.1　参数取值表

	$u_r(t)= 2\sin(2.5\pi t)$	$u_r(t)= 2\sin(2\pi t)$
ρ_0	0.62	0.62
ρ_∞	0.055	0.055
$\bar{\delta}$	1	1
$\underline{\delta}$	1	1
c	0.471	0.471
d	0.456	0.456
α	0.0216	0.0216
ϕ	0.8165	0.83

图 3.2 所示的是在输入为 $u_r(t)= 2\sin(2.5\pi t)$ 时，系统状态 $x_1(t)$，$x_2(t)$ 和参考输入状态 x_{r1}，x_{r2} 的对比图，从图中可以看出，控制系统对状态的跟踪较好，其标量误差在图 3.3 中给出，从图 3.3 中可以明显看出标量误差不论其暂态还是稳态均收敛于规定性能范围内。

图 3.4 所示的是控制器的矢量误差，从误差转换式（3.25）可知闭环系统规定性能控制是根据标量误差设计的，并未对矢量误差做出直接约束，因此从图 3.4 可以明显看出其矢量误差并未被规定性能函数约束；同时若对于纯反馈系统如 $y(t)=x_1(t)$ 类系统来说，转换后的标量误差精度要小于 $e(t)$，这说明转换标量误差虽然可以简化控制器设计，但是却牺牲了部分精度，因此对采用标量误差转换的自适应控制策略采用规定性能控制是十分必要的。

图 3.5 给出了控制器输入 $u(t)$ 和 Hammerstein 系统迟滞非线性环节输出（即线性环节输入）$v(t)$ 的对比，根据图 3.5 可以看出本章所设计的自适应控制策略在未采用逆模型补偿的情况下很好地实现了对该迟滞 Hammerstein 系统的控制。同样的结论也可以从图 3.6～3.9 中得到，此时系统参考输入为 $u_r(t)= 2\sin(2.5\pi t)$。

第 3 章　Backlash-like 迟滞非线性系统预设精度自适应控制

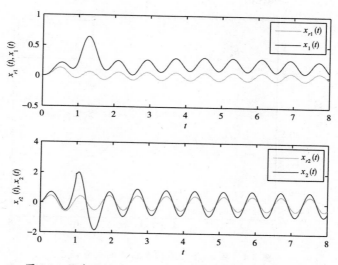

图 3.2　状态 x_1 和 x_{r1}，x_2 和 x_{r2} 在不同输入下的对比图

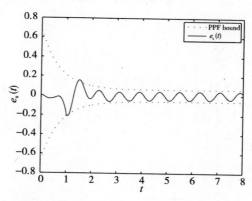

图 3.3　在 $u_r(t)=2\sin(2.5\pi t)$ 下的标量误差

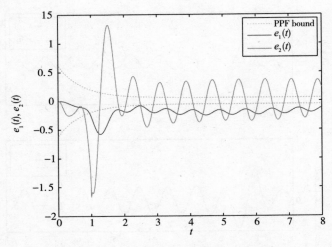

图 3.4 在 $u_r(t)=2\sin(2.5\pi t)$ 下的矢量误差

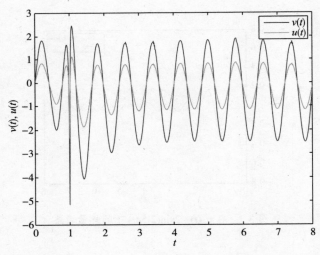

图 3.5 在 $u_r(t)=2\sin(2.5\pi t)$ 下 $u(t)$ 和 $v(t)$ 对比

第 3 章 Backlash-like 迟滞非线性系统预设精度自适应控制

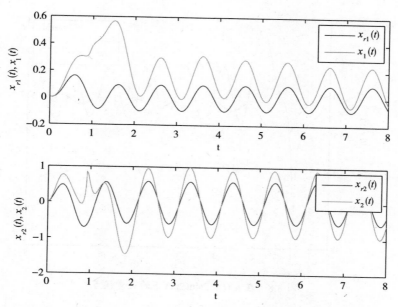

图 3.6 状态 x_1 和 x_{r1}，x_2 和 x_{r2} 在输入 $u_r(t)=2\sin(2\pi t)$ 下的对比图

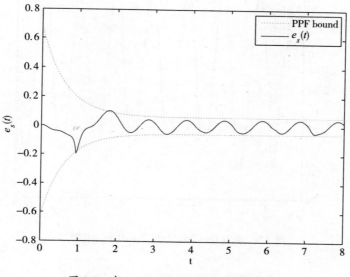

图 3.7 在 $u_r(t)=2\sin(2\pi t)$ 下的标量误差

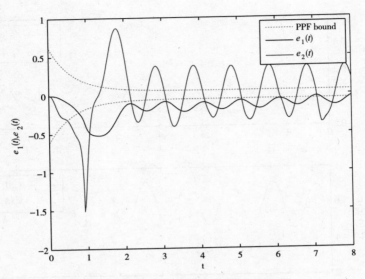

图 3.8 在 $u_r(t)=2\sin(2\pi t)$ 下的矢量误差

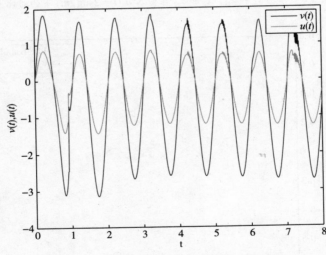

图 3.9 在 $u_r(t)=2\sin(2\pi t)$ 下 $u(t)$ 和 $v(t)$ 对比

通过仿真可知本章所设计自适应控制策略既可以通过标量误差转换简化控制器设计,并可以通过规定性能误差转换使误差收敛于规定性能范围内。因此既保证了闭环控制系统的稳定性和设计简便性,也保证了控制精度。

3.5 结论

本章通过对用 Backlash 类模型描述的迟滞 Hammerstein 系统研究,在文献[6]的研究基础上改进并简化规定性能函数,设计了模型参考自适应控制器控制该 Backlash 类迟滞 Hammerstein 系统。首先为了使控制器进一步简化,将矢量状态误差转化为标量误差,然后根据提出的简化的规定性能函数,将标量误差转化为规定性能误差,在此基础上对该闭环系统设计自适应控制器,并通过 Lambert-W 函数根据 Lyapunov 理论保证了闭环系统的稳定性,这样既简化了控制器设计,又保证了控制性能。仿真结果也验证了该控制策略正确有效。

参考文献

[1] WANG HQ, CHEN B, LIU KF, et al. Adaptive Neural Tracking Control for a Class of Nonstrict-Feedback Sto-chastic Nonlinear Systems with Unknown Backlash-Like Hysteresis[J]. IEEE Transactions on Neural Networks and Learning Systems, 2014,25(5):947-958.

[2] LIU Y, LIN Y. Global adaptive output feedback tracking for a class of non-linear systems with unknown backlash-like hysteresis[J]. IET Control Theory and

Applications, 2014,8(11):927-936.

[3] POULIQUEN M, GIRI F, GEHAN O, et al. Subspace identification for Hammerstein systems with nonparametric input backlash and switch nonlinearities [C]. Conference on Decision and Control, Firenze, Italy, 2013: 4302-4307.

[4] DONG RL, TAN YH. Internal Model Control for Pseudo Hammerstein Systems with Backlash [C]. Robotics, Automation and Mech-atronics, 2008 IEEE Conference on, Chengdu, 2008: 90-94.

[5] BECHLIOULIS CP, ROVITHAKIS GA. Robust Adaptive Control of Feedback Linearizable MIMO Nonlinear Systems with Prescribed Performance [J]. IEEE Transactions on Automatic Control, 2008,53(9):2090-2099.

[6] NA J, CHEN Q, REN XM, et al. Adaptive Prescribed Performance Motion Control of Servo Mechanisms with Friction Compensation [J]. IEEE Transactions on Industrial Electronics, 2014,61(1):486-494.

[7] 赵新龙, 李智, 苏春翌. 基于Bouc-Wen模型的迟滞非线性系统自适应控制 [J]. 系统科学与数学, 2014,34(12):1496-1507.

[8] ZHOU J, WEN CY, ZHANG Y. Adaptive backstepping control of a class of uncertain nonlinear systems with unknown backlash-like hysteresis [J]. IEEE Transactions on Automatic Control, 2004,49(10):1751-1759.

[9] ANURADHA MA, FREDRIK PS, LOH AP. Adaptive control of continuous time systems with convex/concave parametrizetion [J]. Automatica,1998,34(1):33-49.

[10] ABDO B, ARBEL SE, SHTEMPLUCK O, et al. Observation of Bifurcations and

Hysteresis in Nonlinear NbN Superconducting Microwave Resonators [J]. IEEE Transactions on Applied Superconductivity, 2006,16(4):1976-1987.

[11] XU MX, CHEN ZH, XU MQ, et al. Discussion of modified Jiles-Atherton model including dislocations and plastic strain [J]. International Journal of Applied Electromagnetics and Mec-hanics, 2015,47(1):61-73.

[12] TRAPANESE M. Identification of parameters of the Jiles-Atherton model by neural networks [J]. Journal of Applied Physics, 2011(109):07D355.

[13] GYORGY K. On the product Preisach model of hysteresis [J]. Physica B: Condensed Matter, 2000,275(1-3):40-44.

[14] WANG X, REYSETT A, POMMIER BV, et al. A modified Preisach model and its inversion for hysteresis compensation in piezoelectric actuators [J]. Multidiscipline Modeling in Materials and Structures, 2014,10(1):122-142.

[15] MAYERGOYZ I, SERPICO C. Nonlinear diffusion and the Preisach model of hysteresis [J]. Physica B: Condensed Matter, 2000,275(1-3):17-23.

[16] SPANOS PD, CACCIOLA P, REDHORSE J. Random Vibration of SMA Systems via Preisach Formalism [J]. Nonlinear Dynamics, 2004,36(2-4):405-419.

[17] CROSS R, KRASNOSEL'SKII AM, POKROVSKII1AV. A timedepen-dent Preisach model [J]. Physica B: Condensed Matter, 2001,306(1-4):206-210.

第4章　Prandtl–Ishlinskii 迟滞非线性系统的 Hopfield 神经网络辨识与自适应控制

4.1　引言

迟滞非线性系统的辨识问题一直是迟滞非线性系统研究的重点问题[1,2]，但是受到迟滞非线性的多值性等影响，辨识问题一直是迟滞非线性系统研究的难点。本书第 2 章讨论了 Preisach 迟滞模型的辨识问题，提出了下三角矩阵的分段一致辨识，解决了"擦除"的影响。本章针对另外一种迟滞唯象模型 Prandtl-Ishlinskii 迟滞模型（简称 P-I 模型）的辨识问题进行研究。由于迟滞系统严格的非线性，会大大降低系统性能。因此，辨识和处理迟滞非线性就成为一个重要问题。目前对于迟滞辨识的研究取得了一定的成果，但还需要更深入的研究。

一般来说，常见的迟滞模型包括 Jiles-Atherton 模型[3]、Preisach 模型[4]、P-I 模型[5]和 Bouc-Wen 模型[6]。JilesAtherton 模型属于物理模型，其参数与物理材料密切相关。大多数 Jiles-Atherton 迟滞模型都描述使用在了电磁系统。陈鹏和舍甫西克[7]改进了磁滞的 Jiles-Atherton 模型，改进的 Jiles-Atherton 迟滞模型提升了模型微分方程的表达并在晶粒各向同性电工钢或者无晶粒取向电工钢都可以的闭环系统中验证了改进的模型。波普等[8]采用了一种算法辨识磁滞系统的 Jiles-Atherton 模型，该算法利用均方根误差估计迟滞回归曲线以得到辨识结果。

与 Jiles-Atherton 模型不同，其他大多数迟滞模型都是唯象模型。唯象模型由于是数学模型，与实际物理参数联系不大，只选取合适的参数描述物理迟滞现象，因此具有更广泛和更一般的应用性。在唯象模型中，P-I 迟滞模型一直占有重要地位，许多学者进行了深入研究。我们针对 P-I 迟滞非线性系统状态未知的情况下，提出了一种新的扩张状态观测器估计未知状态。将纯反馈系统中未知状态扩张为新的状态，并设计观测器观测状态，在此基础上设计了自适应神经网络控制器控制 P-I 迟滞系统。邹江和谷国迎[9]应用 P-I 迟滞模型描述可用于柔性机器人肌肉中的介电弹性体执行器 (DEA) 的黏弹性迟滞非线性，提出了改进的频率相关 P-I 模型 (MRPIM) 处理非对称的迟滞曲线并应用七个不同频率进行实验验证了方法的可行性。贾奈德和阿尔贾内德[10]利用带有死区函数的频率相关 P-I 模型来表征在相对较高频率或较大幅度输入下具有不对称性和饱和性的磁致伸缩执行器模型，并采用 P-I 逆模型作为前馈补偿器来补偿非对称迟滞非线性。但据笔者所知，使用 Hopfield 神经网络（HNN）辨识 P-I 迟滞模型几乎没有研究。

Hopfield 神经网络是一类反馈神经网络，它可以辨识大量非线性系统的未知参数。阿滕西亚等学者[11]提出了一种新的 Hopfield 神经网络优化算法，用于估计动力系统的时变参数。它放宽了一些通常的统计假设条件，并提高了在某些确定性干扰下的鲁棒性。王志旭和洪昆能[12]提出了一种基于高阶 Hopfield 神经网络（HOHNN），该神经网络具有用于动态系统辨识功能的链接网络（FLN），添加了每个神经元的输入以便使神经网络具有更快的收敛速度和更少的计算负载。笔者[6]在前面的工作中研究了 Buc-Wen 迟滞非线性系统的 Hopfield 神经网络辨识问题。

在本章中，笔者将研究 P-I 非线性系统的辨识和自适应控制问题。

主要创新总结如下：

（1）首先将 P-I 迟滞非线性系统通过正交矩阵变换将其变换为规范形式，提出了合适的能量函数并设计 Hopfield 神经网络辨识系统未知系数。

（2）在辨识的基础上设计自适应控制器，控制 P-I 迟滞非线性系统并保证闭环系统的稳定性。

4.2　问题描述

4.2.1　Prandtl-Ishlinskii 迟滞系统描述

考虑一类具有 P-I 迟滞模型的系统定义如下：

$$\begin{cases} \dot{x} = Ax + B\psi(u) \\ y = Cx \end{cases} \quad (4.1)$$

其中，$A \in \mathbb{R}^{n \times n}$，$B \in \mathbb{R}^n$ 为状态向量的未知系数，$\psi(u)$ 表示迟滞非线性。迟滞 P-I 模型定义如下：

$$\psi(u) = \upsilon_0 u(t) + \int_0^R \upsilon(r) \Omega_r[u](t) \mathrm{d}r \quad (4.2)$$

其中，υ_0 表示 P-I 模型的初始值，且 $\upsilon_0 > 0$ 为正常数。当 $\upsilon_0 > 0$ 时，可积密度函数 $\upsilon(r)$ 满足 $\int_0^R r\upsilon(r) \mathrm{d}r < \infty$，$\Omega_r[u](t)$ 表示 Play 或 Stop 算子。在本章中，选择 play 算子

来定义 $\Omega_r[u](t)$，play 算子如下所示[13]：

$$\begin{cases} \varpi_0(0) = \Omega_r[u](0) = g_r(u(0), \varpi_{m-1}(0)) \\ \varpi_m(t) = \Omega_r[u](t) = g_r(u(t), \varpi_m(t_i)) \end{cases} \quad (4.3)$$

其中，$g_r(u, \varpi_m) = \max\{(u-r), \min\{(u+r, \varpi_m)\}\}$，$r \geqslant 0$，$t_i \leqslant t \leqslant t_{i+1}$。$\varpi_{m-1}$ 为给出的初始值。本章考虑系统状态已知，状态参数未知，情况下辨识状态参数及设计自适应控制器问题。

本章目标为：

（1）将 P-I 迟滞非线性系统通过正交矩阵变换为规范形式；

（2）设计 Hopfield 神经网络辨识该 P-I 迟滞系统；

（3）设计自适应控制器控制闭环迟滞系统，并保证系统稳定性。

4.2.2 系统变换

考虑系统（4.1），给出假设如下：

假设 4.1 （A1）系统（4.1）是可控和可观的。

（A2）传递函数 $G(s) = C(SI-A)^{-1}B$ 满秩。

（A3）增益 $K_g = CB$ 已知。

在假设 4.1 下，由于系统可控可观且增益已知，则可将系统（4.1）基于假设 4.1 转换为规范的状态空间形式。通过非奇异线性状态转换[1]，以下内容成立：

$$g_r(u, \varpi_m) = \max\{(u-r), \min\{(u+r), \varpi_m\}\}$$

$$\begin{cases} \begin{bmatrix} w \\ z \end{bmatrix} = \begin{bmatrix} B^\perp \\ C \end{bmatrix} x \\ B^\perp B = 0 \end{cases} \quad (4.4)$$

因此，将 P-I 迟滞系统 (4.1) 转换为规范状态空间形式如下：

$$\begin{cases} \dot{w} = A_{11}w + A_{12}z \\ \dot{z} = A_{21}w + A_{22}z + K_g\psi(u) \end{cases} \quad (4.5)$$

其中，$\omega \in \mathbb{R}^{n-m}$，$z \in \mathbb{R}^m$。

不失一般性，我们选择 z 的维数 $m=1$ 以简化控制设计。则规范形式（4.5）可以改写为：

$$\begin{bmatrix} \dot{w} \\ \dot{z} \end{bmatrix} = \begin{bmatrix} A_{11} & A_{12} \\ A_{21} & A_{22} \end{bmatrix} \begin{bmatrix} w \\ z \end{bmatrix} + \begin{bmatrix} 0 \\ K_g \end{bmatrix} \psi(u) = \bar{A} \begin{bmatrix} w \\ z \end{bmatrix} + \bar{B}\psi(u) \quad (4.6)$$

4.3 Hopfield 神经网络辨识

本节将设计一个 Hopfield 神经网络来估计由参数矩阵 A 变换的未知参数矩阵 $\bar{A}$。首先，引入 Hopfield 神经网络并将其重写为矩阵表示形式。然后，定义合适的误差，并通过该误差提出能量函数 E，并定义未知 $\bar{A}$ 用于辨识。在 Hopfield 神经网络下，定义另一个能量函数 E_n，令它等于能量 E，通过解等式 $E=E_n$ 获得未知的 $\bar{A}$。

4.3.1 Hopfield 神经网络

首先定义 Hopfield 神经网络，Hopfield 神经网络第 i 个神经元可以描述如下：

$$\frac{du_i}{dt} = -\frac{1}{D_i}\left(\frac{1}{S_i}u_i(t) + \sum_j W_{ij}f_i(u_j(t)) + L_i\right) \quad (4.7)$$

其中，u_i 是Hopfield神经元的输入，D_i、S_i、W_{ij}、L_i分别表示对应的电容、电阻、j神经元与i神经元连接的权重、i神经元外部输入。f_i表示一个严格递增的、有界的、非线性连续函数。本章假设$R_i=\infty$和$C_i=1$,此外,考虑到Hopfield神经网络有M个神经元,可得:

$$\frac{\mathrm{d}u_i}{\mathrm{d}t} = -\left(\sum_{j=1}^{M} W_{ij} Q_j(t) + L_i\right) \quad (4.8)$$

其中,Q_j代表神经元j的输出,由下式给出

$$Q_j(t) = \rho \tanh\left(\frac{u_i(t)}{\eta}\right) \quad (4.9)$$

其中,$\rho, \eta > 0$。

因此Hopfield神经网络矩阵形式可表示如下:

$$\frac{\mathrm{d}u}{\mathrm{d}t} = -(W(t)Q(t) + L(t))$$

$$Q(t) = \rho \tanh\left(\frac{u(t)}{\eta}\right) \quad (4.10)$$

4.3.2 Hopfield神经网络辨识

为了辨识未知参数$\bar{A}$,方程(4.6)表示为

$$\bar{x} = \bar{A}\bar{x} + \bar{B}\psi(u) \quad (4.11)$$

假设参考系统为

$$\hat{x} = A_p \bar{x} + \bar{B}\psi(u) \quad (4.12)$$

则,辨识误差可以定义为

$$e_x = x - \hat{x} \quad (4.13)$$

有下列定理成立：

定理 4.1　若有 P-I 迟滞非线性系统（4.1）（4.2）和（4.3），将其规范状态空间形式通过正交矩阵变换（4.4）（4.5）和（4.6）变换为（4.11），Hopfield 神经网络选择为（4.10），那么，设计合适的 Hopfield 神经网络辨识 $\bar{A}$，且可用（4.12）估计迟滞系统（4.11）。

证明　考虑辨识误差（4.13），则误差的导数可以由（4.11）和（4.12）推导出：

$$\dot{e}_x = \dot{\bar{x}} - \dot{\hat{x}} = \dot{\bar{x}} - (A_p \bar{x} + \bar{B}\psi(u)) = \dot{\bar{x}} - \left(A_p \bar{x} + \bar{B}\left(\upsilon_0 u(t) + \int_0^R \upsilon(r)\Omega_r[u](t)\mathrm{d}r \right) \right) \quad (4.14)$$

选择能量函数 E 如下所示：

$$E = \frac{1}{2}\dot{e}_x^{\mathrm{T}}\dot{e}_x \quad (4.15)$$

考虑到（4.14），可以得到：

$$\begin{aligned}
E &= \frac{1}{2}\left(\dot{\bar{x}} - A_p\bar{x} - \bar{B}\left(\upsilon_0 u(t) + \int_0^R \upsilon(r)\Omega_r[u](t)\mathrm{d}r\right)\right)^{\mathrm{T}} \\
&\quad \left(\dot{\bar{x}} - A_p\bar{x} - \bar{B}\left(\upsilon_0 u(t) + \int_0^R \upsilon(r)\Omega_r[u](t)\mathrm{d}r\right)\right) \\
&= \frac{1}{2}\Big(\dot{\bar{x}}^{\mathrm{T}}\dot{\bar{x}} - \dot{\bar{x}}^{\mathrm{T}}A_p\bar{x} - \upsilon_0 u(t)\dot{\bar{x}}^{\mathrm{T}}\bar{B} - \dot{\bar{x}}^{\mathrm{T}}\bar{B}\int_0^R \upsilon(r)\Omega_r[u](t)\mathrm{d}r \\
&\quad -\bar{x}^{\mathrm{T}}A_p^{\mathrm{T}}\dot{\bar{x}} + \bar{x}^{\mathrm{T}}A_p^{\mathrm{T}}A_p\bar{x} + \upsilon_0 u(t)\bar{x}^{\mathrm{T}}A_p^{\mathrm{T}}\bar{B} + \bar{x}^{\mathrm{T}}A_p^{\mathrm{T}}\bar{B}\int_0^R \upsilon(r)\Omega_r[u](t)\mathrm{d}r \\
&\quad -\upsilon_0 u(t)\bar{B}^{\mathrm{T}}\dot{\bar{x}} + \upsilon_0 u(t)\bar{B}^{\mathrm{T}}A_p\bar{x} + \upsilon_0^2 u^2(t)\bar{B}^{\mathrm{T}}\bar{B} \\
&\quad +\upsilon_0 u(t)\bar{B}^{\mathrm{T}}\bar{B}\int_0^R \upsilon(r)\Omega_r[u](t)\mathrm{d}r - \bar{B}^{\mathrm{T}}\dot{\bar{x}}\int_0^R \upsilon(r)\Omega_r[u](t)\mathrm{d}r \\
&\quad +\bar{B}^{\mathrm{T}}A_p\bar{x}\int_0^R \upsilon(r)\Omega_r[u](t)\mathrm{d}r + \upsilon_0 u(t)\bar{B}^{\mathrm{T}}\bar{B}\int_0^R \upsilon(r)\Omega_r[u](t)\mathrm{d}r \\
&\quad +\bar{B}^{\mathrm{T}}\bar{B}\left(\int_0^R \upsilon(r)\Omega_r[u](t)\mathrm{d}r\right)^2\Big)
\end{aligned} \quad (4.16)$$

为了估计未知参数 $\bar{A}$, 定义一个向量为:

$$V = [\bar{A}, \bar{B}] = [a_{11}, a_{12}, \cdots, a_{nn}, 0, \cdots, 1] \quad (4.17)$$

则基于 Hopfield 神经网络, 定义另一个能量函数 E_n 如下所示:

$$E_n = \frac{1}{2}WV - LV \quad (4.18)$$

令 $E = E_n$, 可得如下方程:

$$WV = \dot{\overline{x}}^{\text{T}}\dot{\overline{x}} + \overline{x}^{\text{T}}A_p^{\text{T}}A_p\overline{x} + \upsilon_0 u(t)\overline{x}^{\text{T}}A_P^{\text{T}}\overline{B} + \overline{x}^{\text{T}}A_P^{\text{T}}\overline{B}\int_0^R \upsilon(r)\Omega_r[u](t)\text{d}r$$

$$+ \upsilon_0 u(t)\overline{B}^{\text{T}}A_p\overline{x} + \upsilon_0^2 u^2(t)\overline{B}^{\text{T}}\overline{B} + \upsilon_0 u(t)\overline{B}^{\text{T}}\overline{B}\int_0^R \upsilon(r)\Omega_r[u](t)\text{d}r$$

$$+ \overline{B}^{\text{T}}A_p\overline{x}\int_0^R \upsilon(r)\Omega_r[u](t)\text{d}r + \upsilon_0 u(t)\overline{B}^{\text{T}}\overline{B}\int_0^R \upsilon(r)\Omega_r[u](t)\text{d}r$$

$$+ \overline{B}^{\text{T}}\overline{B}\left(\int_0^R \upsilon(r)\Omega_r[u](t)\text{d}r\right)^2$$

$$LV = \dot{\overline{x}}A_p\overline{x} + \upsilon_0 u(t)\dot{\overline{x}}^{\text{T}}\overline{B} + \dot{\overline{x}}^{\text{T}}\overline{B}\int_0^R \upsilon(r)\Omega_r[u](t)\text{d}r + \overline{x}^{\text{T}}A_P^{\text{T}}\dot{\overline{x}}$$

$$+ \upsilon_0 u(t)\overline{B}^{\text{T}}\dot{\overline{x}} + \overline{B}^{\text{T}}\dot{\overline{x}}\int_0^R \upsilon(r)\Omega_r[u](t)\text{d}r \tag{4.19}$$

通过计算方程（4.19），可获取参数 V，由此进一步得到 $\overline{A}$。

证毕。

4.4 自适应控制器设计

根据上节 Hopfield 神经网络辨识算法得到 $\overline{A}$ 后，可设计迟滞系统自适应控制器。考虑系统（4.1）转换为系统（4.11），重新写为

$$\begin{cases} \dot{\overline{x}} = \overline{A}\overline{x} + \overline{B}\psi(u) \\ \psi(u) = \upsilon_0 u(t) + \int_0^R \upsilon(r)\Omega_r[u](t)\text{d}r \end{cases} \tag{4.20}$$

则存在正定对称矩阵 P 使得下式成立：

$$\overline{A}^{\text{T}}P + P\overline{A} = -Q \tag{4.21}$$

其中，Q 也为正定对称矩阵。

控制器设计为

$$u(t)=-\frac{1}{v_0}\int_0^R \upsilon(r)\varOmega_r[u](t)\mathrm{d}r \qquad (4.22)$$

选择李雅普诺夫函数为

$$V_c=\frac{1}{2}\bar{x}^\mathrm{T}P\bar{x} \qquad (4.23)$$

则有：

$$\begin{aligned}\dot{V}_c&=\frac{1}{2}\dot{\bar{x}}^\mathrm{T}P\bar{x}+\frac{1}{2}\bar{x}^\mathrm{T}P\dot{\bar{x}}\\&=\frac{1}{2}\left(\bar{x}^\mathrm{T}\bar{A}^\mathrm{T}+\bar{B}^\mathrm{T}\psi(u)\right)P\bar{x}+\frac{1}{2}\bar{x}^\mathrm{T}P\left(\bar{A}\bar{x}+\bar{B}\psi(u)\right)\\&=-\frac{1}{2}\bar{x}^\mathrm{T}Q\bar{x}+\bar{B}^T P\bar{x}\left(v_0 u(t)+\int_0^R \upsilon(r)\varOmega_r[u](t)\mathrm{d}r\right)\end{aligned} \qquad (4.24)$$

将控制器（4.22）带入（4.24）可得：

$$\dot{V}_c \leq 0 \qquad (4.25)$$

因此根据李雅普诺夫稳定性理论，闭环控制系统稳定。

4.5 仿真

本节将验证所提出的Hopfield神经网络辨识方法和自适应控制方法。首先，考虑迟滞非线性系统（4.11），其中

$$\bar{A}=\begin{bmatrix}0 & 0.3\\-3.2 & -2\end{bmatrix},\bar{B}=\begin{bmatrix}0\\1\end{bmatrix} \qquad (4.26)$$

$\phi(u)$ 被描述为（4.2），其中 $v_0=2.5$，$\upsilon(r)=0.053\mathrm{e}^{-0.062(r-1)^2}$，$R=50$. 考虑到辨识模型（4.12），估计参数 A_p 可以定义为：

$$A_p = \begin{bmatrix} a_{11} & a_{12} \\ a_{21} & a_{22} \end{bmatrix} \tag{4.27}$$

且定义 $V=[a_{11}, a_{12}, a_{21}, a_{22}, 0.1]^T$。

根据式（4.19）和 $E=E_n$，可求出 Hopfield 神经网络的权重 W 和 I 为：

$$W = \begin{bmatrix} \bar{x}_1^2 & \bar{x}_1\bar{x}_2 & 0 & 0 & \bar{x}_1 u & 0 \\ \bar{x}_2\bar{x}_1 & \bar{x}_2^2 & 0 & 0 & \bar{x}_2 u & 0 \\ 0 & 0 & \bar{x}_1^2 & \bar{x}_1\bar{x}_2 & 0 & \bar{x}_1 u \\ 0 & 0 & \bar{x}_2\bar{x}_1 & \bar{x}_2^2 & 0 & \bar{x}_2 u \\ u\bar{x}_1 & u\bar{x}_2 & 0 & 0 & u^2 & 0 \\ 0 & 0 & u\bar{x}_1 & u\bar{x}_2 & 0 & u^2 \end{bmatrix}, \quad I = \begin{bmatrix} \bar{x}_1\bar{x}_1 \\ \bar{x}_2\bar{x}_1 \\ \bar{x}_1\bar{x}_2 \\ \bar{x}_2\bar{x}_2 \\ u\bar{x}_1 \\ u\bar{x}_2 \end{bmatrix} \tag{4.28}$$

然后，根据（4.26）（4.27）和（4.28）进行模拟以验证所提出的 Hopfield 神经网络辨识方法和自适应控制方法的有效性。辨识结果和辨识误差如图 4.1 和图 4.2 所示。图 4.1 清楚地说明了所提出的 Hopfield 神经网络可以精确地估计未知参数 $\bar{A}$。图 4.2 表明误差收敛到一个很小的零邻域。此外，a_{11}，a_{12}，a_{21}，a_{22} 的平均绝对误差（MAE）如表 4.1 所示。

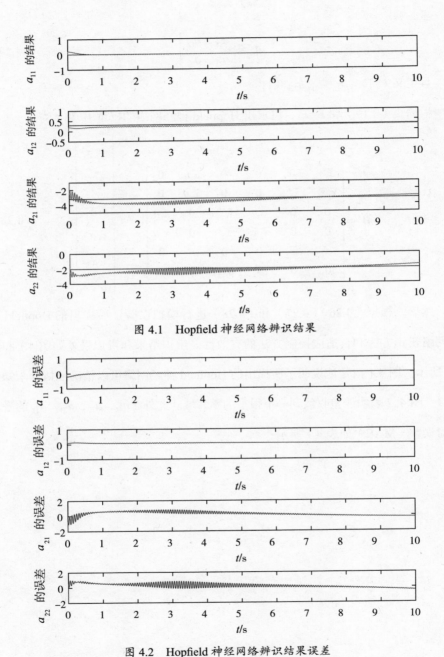

图 4.1 Hopfield 神经网络辨识结果

图 4.2 Hopfield 神经网络辨识结果误差

第 4 章 Prandtl–Ishlinskii 迟滞非线性系统的 Hopfield 神经网络辨识与自适应控制

表 4.1 辨识平均绝对误差

a_{11}	a_{12}	a_{21}	a_{22}
0.0076	0.0471	0.0613	0.5011

从表 4.1 可以看出，对于用 P-I 迟滞模型描述的迟滞非线性系统，Hopfield 神经网络可以精确的辨识其参数，仿真结果表明本章所提出的辨识方法是有效的。图 4.3 和图 4.4 是所提出自适应控制的仿真结果和误差。从图 4.3 可以看出自适应控制很好地控制了 P-I 迟滞非线性系统。图 4.4 表明控制误差很小，控制器平均绝对误差对状态 $\bar{x}_1, \bar{x}_2$ 来说分别为 0.042 和 0.067，达到了理想的控制效果。因此，仿真证明所采用的自适应控制策略可靠有效。

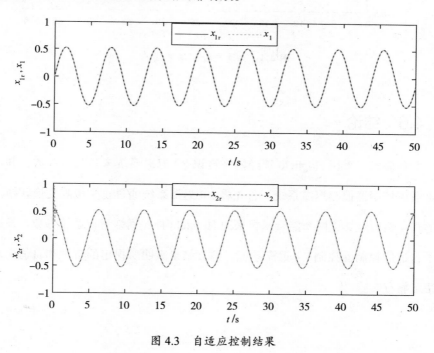

图 4.3 自适应控制结果

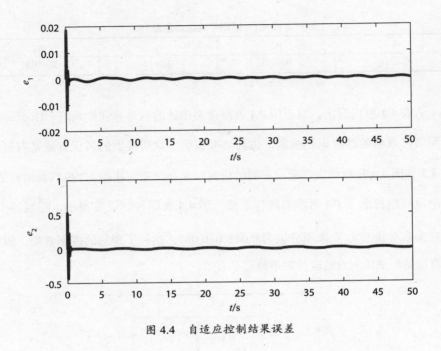

图 4.4 自适应控制结果误差

4.6 结论

本章提出了采用 Hopfield 神经网络辨识 P-I 迟滞系统未知状态参数，并设计了自适应控制器控制迟滞系统。首先将 P-I 迟滞系统通过正交矩阵变换转换为规范形式，然后定义了两个能量函数采用 Hopfield 神经网络辨识系统参数。最后设计了自适应控制器控制 P-I 迟滞系统，仿真结果表明所提出的辨识和自适应控制方法可靠有效。

参考文献

[1] NA J, CHEN AS, HERRMANN G, et al. Vehicle engine torque estimation via unknown input observer and adaptive parameter estimation [J]. IEEE Transactions on Vehicular Technology, 2018, 67(1):409–422.

[2] WANG SB, NA J, REN XM. Rise-based asymptotic prescribed performance tracking control of nonlinear servo mechanisms [J]. IEEE Transactions on Systems, Man, and Cyber-netics, 2017, (99): 1–12.

[3] ZHANG DH, JIA MF, LIU Y, et al. Comprehensive improvement of temperature dependent Jiles-Atherton model utilizing variable model parameters [J]. IEEE Transactions on Magnetics. 2018, 54(3):1–4.

[4] GAO XH, REN XM, ZHU CS, et al. Identification and control for ham-merstein systems with hysteresis non-linearity [J]. IET Control Theory & Application, 2015, 9(13): 1935–1947.

[5] GAO XH, LIU RG, SUN B, et al. Neural Network Adaptive Control for Hysteresis Hammerstein System [C]. Proceedings of 2017 Chinese Intelligent Systems Conference, Mudanjiang, 2018(459): 259–269.

[6] GAO XH, SUN B. Identification for Bouc-Wen hysteresis system with Hopfield network [C]. 2017 9th International Conference on Modelling, Identification and Control (ICMIC), 2017: 248–253.

[7] CHENG P, SZEWCZYK R. Modified Description of Magnetic Hysteresis in Jiles-Atherton Model [C] . Conference on Automation 2018, Warsaw, Poland, 2018(743): 648–654.

[8] POP N, CALTUN O. Jiles-Atherton magnetic hysteresis parameters iden-tification [J] . Acta Physica Polonica, 2011, 120(3): 491–496.

[9] ZOU J, GU GY. Modeling the viscoelastic hysteresis of dielectric elastomer actuators with a modified rate-dependent Prandtl-Ishlinskii model [J] . Polymers, 2018, 10(5):525.

[10] JANAIDEH MA, ALJANAIDEH O. Further results on open-loop compensation of rate-dependent hysteresis in a magneto strictive actuator with the Prandtl-Ishlinskii model [J] . Mechanical Systems & Signal Processing, 2018(104): 835–850.

[11] ATENCIA M, JOYA G, SANDOVAL F. Identification of noisy dynamical systems with parameter estimation based on hopfield neural networks [J] . Neurocomputing, 2013(121): 14–24.

[12] WANG CH, HUNG KN. Dynamic system identification using high-order hopfield-based neural network (HOHNN) [J] . Asian Journal of Control, 2012, 14(6): 1553–1566.

[13] LIU S, SU CY. Inverse error analysis and adaptive output feedback control of uncertain systems preceded with hysteresis actuators [J] . IET Control Theory Application, 2014, 8(17): 1824–1832.

[14] VIDAL PVNM, NUNES EVL, HSU L. Output-feedback multi-variable global variable gain super-twisting algorithm [J]. IEEE Transactions on Automatic Control, 2017, 62(6): 2999–3005.

第 5 章　Bouc-Wen 迟滞非线性系统扩张状态观测器与回声神经网络滑模控制

5.1　引言

前几章针对不同的迟滞模型进行了辨识和控制策略研究，本章将对 Bouc-Wen 迟滞模型描述的迟滞非线性系统进行研究。

本章以驱动电动汽车的永磁同步电机（PMSM）模型为基础研究迟滞非线性系统的控制。大多数电动汽车采用永磁同步电机驱动，但是电动汽车系统中不仅包括永磁同步电机，还具有许多其他子系统。那靖等学者[1]讨论了汽车不同部分的建模问题，并设计输入观测器和自适应估计器估计未知参数。为了获得更好的控制精度，文献[2]采用规定的性能函数精确控制车辆悬架的精度。王树波等学者[3,4]采用漏斗形控制函数并在此基础上设计了一种自适应控制器来精确控制永磁同步电机模型。黄英博等学者[5]提出了一种不需要函数逼近器（例如，神经网络和模糊逻辑系统）的车辆悬架自由近似的自适应控制策略。不仅如此，由于车辆的悬架系统通常存在输入延迟情况，那靖等学者[6]提出了一种考虑输入延迟的有限时间自适应模糊控制策略控制车辆非线性主动悬架系统。文献[7]采用了基于扩展卡尔曼滤波器的最小模型误差跟踪控制自动驾驶汽车系统，该控制策略考虑了实际车辆系统中的输入饱和问题。与以上文献不同，李博源等学者[8]用 MP 算法处理非线性轮胎模型。而文献[9,10]用磁流变（MR）可变刚度和阻尼模型研究车辆悬架系统。其中阻尼器模型由两个 Bouc-Wen 迟滞模型描述。本章与文献[9]类似，也采用 Bouc-Wen 迟滞模型描述永磁同步电机的迟滞非线性。

虽然上述文献研究了车辆的全部模型或部分模型（子模型），但是这些研究中既包含了纯电动汽车系统，也包含了混合动力汽车系统。在纯电动汽车系统中，电机模型的研究首当其冲。目前电动汽车电机系统多种多样，但是永磁同步电机是最重要的一类。

永磁同步电机模型的研究已经进行了多年，在不同条件下有不同的建模方法，也获得了不同的成果。骆光照等学者[11]提出了一种永磁同步电机的场路耦合参数自适应建模方法，它结合了数学模型和磁场模型两方面的优点。特别是针对永磁材料中切实存在的磁滞损耗问题，许多学者进行了深入研究。埃戈罗夫[12]等学者采用静态历史相关迟滞模型（HDHM）研究永磁体的迟滞损耗问题，并将其应用于转子表面磁体采用铁氧磁体的永磁同步电机系统。但是文献[13]则提出了一种带有空间矢量脉冲调节的零序电流迟滞控制器控制开放式绕组永磁同步电机系统。

尽管永磁同步电机已经被广泛研究并取得大量成果[14,15]，但是迟滞模型问题、状态观测等问题仍然是难以确切解决的难题。文献[16-19]采用 U 模型理论研究了状态观测问题。但是考虑到在许多情况下，实际系统难以直接获得模型所有状态，因此，永磁同步电机迟滞系统的状态观测仍需要进一步深入研究。通常情况下，观测器可用于估计受控系统中的两类未知数：干扰和状态[4,20-24]。多年以来，许多不同的观测器已经被广泛应用状态观测或扰动观测。在这些观测器中，扩张状态观测器（ESO）是近年来表现非常优秀的观测器之一，取得了很好的观测效果。

扩张状态观测器最早由韩京清[25]提出，之后众多优秀学者对此产生了浓厚的研究兴趣。薛文超等学者[26]提出了一种基于扩张状态观测器的自抗扰控制（ADRC）来处理系统的不确定性，其中扩张状态观测器的增益可以自动及时地调整以减少估计误差，同时薛文超等学者[27]还证明了扩张状态观测器可以通过提高增益来增加估计精度。陈强等学者[28]提出了一种扩张状态观测器来估计高阶非线性系统未知状态和扰动，以简化并利于控制器设计。孙国法等学者[29]针对多电机伺服系统，使用扩张状态观测器估计伺服电机的未知速度，而王树波等学者[30]也针对伺服电机系统进行研究，但与文献[29]不同，扩张状态观测器被用来观测控制系统中的未知动态（例如摩擦和扰动）。

总体说来，大多数扩张状态观测器被用于估计"总扰动"，因此被视为控制系统的扰动观测器。在本章中，我们将改进扩张状态观测器的结构以观测系统状态，改进后的扩张状态观测器采用高增益参数，可以快速估计未知系统状态，精度更高。为了简化扩张状态观测器设计并保证其收敛性，永磁同步电机迟滞模型的迟滞非线性和其他未知部分作为"一般干扰"将由回声神经网络（ESN）处理。

回声神经网络因其较少的节点和较少的计算需求而被用于许多非线性系统。

陈强等学者[31]为一类受约束的纯反馈非线性系统设计了一种自适应回声神经网络控制器控制系统,而孙国法等学者[32]则针对多输入多输出(MIMO)非线性系统设计了一种改进的回声神经网络动态面控制器,都取得了很好的结果。此外,王树波等学者[33,34]提出了具有回声神经网络的非线性伺服电机系统的规定性能跟踪控制,分别采用鲁棒自适应控制和动态面控制并通过实验验证。同时,王树波等学者[35]针对双质量系统设计了回声神经网络规定性能动态面控制,并考虑了伺服系统间隙的影响。本章将采用回声神经网络处理永磁同步电机系统的未知迟滞非线性并简化扩张状态观测器设计。

滑模控制(SMC)是近年来广泛研究的优秀控制策略。在第2章中,笔者[36]针对转台伺服系统设计了一个复合控制,它由离散逆模型控制器和离散自适应滑模控制器组成来处理Hammerstein系统。为了抑制滑模控制中的"抖震"问题,笔者以及陶亮等学者[36,37]采用双曲函数$\tanh(\cdot)$代替符号函数$\text{sign}(\cdot)$有效降低了"抖震"的影响。赵威等学者[38]将滑模控制应用于多电机驱动伺服系统,陈强等学者[39]对具有不确定性的航天器系统采用滑模控制。本章在回声神经网络估计和扩张状态观测器观测系统状态的基础上采用自适应滑模控制器控制永磁同步电机迟滞系统。

本章首先考虑之前大多永磁同步电机模型研究中被忽略的迟滞非线性的影响,建立了一个考虑磁滞的新的改进的永磁同步电机迟滞系统模型。其次,将改进的永磁同步电机迟滞模型转换为规范形式以利于控制器设计。由于系统状态无法全部直接测得,提出了扩张状态观测器观测未知系统状态,并采用回声神经网络估计系统未知迟滞非线性和扰动。在此基础上设计了自适应滑模控制器控制闭环系统并保证闭环系统的稳定性。

本章创新点如下：

（1）考虑迟滞非线性的影响，改进了永磁同步电机模型以适合实际电动车电机系统，并将改进后的永磁同步电机模型转换为规范形式，以利于简化扩张状态观测器和自适应滑模控制器设计。

（2）与大多数作为扰动观测器用的扩张状态观测器不同，调整了扩张状态观测器的结构用来观测改进后的永磁同步电机模型的系统状态，将扩张状态观测器作为状态观测器使用，并采用回声神经网络补偿未知迟滞非线性和扰动，从而简化了观测器设计。

（3）针对改进的永磁同步电机迟滞系统和扩张状态观测结果，设计了自适应滑模控制器控制迟滞非线性系统，并利用连续函数 tanh（·）代替不连续函数符号 sign（·）来抑制滑模控制的"抖震"问题。

5.2　问题表述

5.2.1　问题描述

本节研究具有迟滞非线性的应用于电动汽车的永磁同步电机模型。根据参考文献［40］，永磁同步电动机 $d\text{-}q$ 模型可以描述如下：

$$\begin{cases}\dfrac{\mathrm{d}i_q}{\mathrm{d}t}=-\dfrac{R}{L_q}i_q+\dfrac{\omega L_d}{L_q}i_d-\dfrac{K_l}{L_q}\omega+\dfrac{1}{L_q}v_q\\ \dfrac{\mathrm{d}i_d}{\mathrm{d}t}=-\dfrac{R}{L_d}i_d-\dfrac{\omega L_q}{L_d}i_q+\dfrac{1}{L_d}v_d\\ T=K_t i_q+K_l\left(L_d-L_q\right)i_d i_q\end{cases}\quad(5.1)$$

其中，i_d，i_q，v_d和v_q分别代表电机d-q轴的电流和电压；R和L_d，L_q分别表示电机电阻和d-q轴电感。K_l和K_t分别是电动机感应的电压常数和电动机转矩常数；T表示传递扭矩；ω表示电动机角速度。

从（5.1）可以看出，电动机模型参数取值为常数，这样就忽略了磁滞损耗、交叉耦合、涡流损耗等。此外，根据参考文献［40］的讨论，由于转子齿的数量足够多，因此可以假定$L_d\approx L_q$。考虑到先前模型中忽略因素的影响，尤其是磁滞损耗，将模型（5.1）修改为

$$\begin{cases}\dfrac{\mathrm{d}i_q}{\mathrm{d}t}=-\dfrac{R}{L_q}i_q+\dfrac{\omega L_d}{L_q}i_d-\dfrac{K_l}{L_q}\omega+\dfrac{1}{L_q}v_q+f_1(i_q,i_d)\\ \dfrac{\mathrm{d}i_d}{\mathrm{d}t}=-\dfrac{R}{L_d}i_d-\dfrac{\omega L_q}{L_d}i_q+\dfrac{1}{L_d}v_d+f_2(i_q,i_d)\\ T\approx K_t i_q\end{cases}\quad(5.2)$$

其中，$f_1(i_q,i_d)$，$f_2(i_q,i_d)$和$g_1(i_q,i_d)$是未知的非线性平滑函数，它们表示（5.1）中诸如磁滞损耗之类的被忽略的非线性因素。将模型（5.2）改写为状态空间形式如下：

$$\begin{cases} \dfrac{\mathrm{d}}{\mathrm{d}t}\begin{bmatrix} i_q \\ i_d \end{bmatrix} = \begin{bmatrix} -\dfrac{R}{L_q} & \dfrac{\omega L_d}{L_q} \\ -\dfrac{\omega L_q}{L_d} & -\dfrac{R}{L_d} \end{bmatrix}\begin{bmatrix} i_q \\ i_d \end{bmatrix} + \begin{bmatrix} \dfrac{1}{L_q} \\ \dfrac{1}{L_d} \end{bmatrix}\begin{bmatrix} v_q \\ v_d \end{bmatrix} + \begin{bmatrix} f_1(i_q,i_d) \\ f_2(i_q,i_d) \end{bmatrix} \\ \dfrac{T}{K_t} = i_q \end{cases} \quad (5.3)$$

其中，

$$\boldsymbol{x} = \begin{bmatrix} x_1 \\ x_2 \end{bmatrix} = \begin{bmatrix} i_q \\ i_d \end{bmatrix}, \quad \boldsymbol{A} = \begin{bmatrix} a_{11} & a_{12} \\ a_{21} & a_{22} \end{bmatrix} = \begin{bmatrix} -\dfrac{R}{L_q} & \dfrac{\omega L_d}{L_q} \\ -\dfrac{\omega L_q}{L_d} & -\dfrac{R}{L_d} \end{bmatrix},$$

$$\boldsymbol{B} = \begin{bmatrix} b_1 \\ b_2 \end{bmatrix} = \begin{bmatrix} \dfrac{1}{L_q} \\ \dfrac{1}{L_d} \end{bmatrix}, \quad \boldsymbol{f}(x) = \begin{bmatrix} f_1(x) \\ f_2(x) \end{bmatrix} = \begin{bmatrix} f_1(i_q,i_d) \\ f_2(i_q,i_d) \end{bmatrix}. \quad (5.4)$$

本章目标：

（1）将模型（5.4）转化为规范形式以利于观测器和控制器设计；

（2）设计扩张状态观测器观测未知状态；

（3）设计回声神经网络估计系统迟滞非线性和其他未知扰动；

（4）设计自适应滑模控制器控制迟滞非线性闭环系统并保证稳定性。

5.2.2 模型转换

大多数观测器和控制器都要求系统为规范形式，为了有利于观测器和控制器设计，将使用一组新的状态变量将系统模型（5.2）转换为规范形式。新的系统状态定义为

$$\begin{cases} z_1 = x_1, \\ z_2 = \dot{z}_1. \end{cases} \quad (5.5)$$

根据等式 $z_2 = \dot{z}_1 = \ddot{x}_1 = -\left(\dfrac{R}{L_q}\right)i_q + \left(\dfrac{\omega L_d}{L_q}\right)i_d - \left(\dfrac{K_t}{L_q}\right)\omega + \left(\dfrac{1}{L_d}\right)v_q + f_1(i_q, i_d)$

得到：

$$\begin{aligned}\dot{z}_2 &= \dot{z}_1 = -\frac{R}{L_q}z_2 + \frac{\omega L_d}{L_q}\left(-\frac{R}{L_d}i_d - \frac{\omega L_q}{L_d}i_q + \frac{1}{L_d}v_d + f_2(x)\right) + \dot{f}_1(x) \\ &= \underbrace{-\omega^2}_{\bar{a}_1}z_1 - \underbrace{\frac{R}{L_q}}_{\bar{a}_2}z_2 + \underbrace{\frac{\omega}{L_q}}_{b}\underbrace{v_d}_{u} - \underbrace{\frac{\omega R}{L_q}i_d - \frac{\omega L_d}{L_q}f_2(x) + \dot{f}_1(x)}_{F(x)}\end{aligned} \quad (5.6)$$

在本章中，采用 Bouc-Wen 迟滞模型来描述汽车永磁同步电机 d-q 模型的磁滞损耗。给出 Bouc-Wen 模型如下：

$$f(x) = \varsigma_1 u + \varsigma_2 \chi, \quad \dot{\chi} = \varpi_0 \dot{u} - \varpi_1 |\dot{u}||\chi|^{h-1}\chi - \varpi_2 \dot{u}|\chi|^h, \quad (5.7)$$

其中，

$\varpi_0 > 0$，$\varpi_1 > |\varpi_2|$，$h > 1, \chi(0) = 0, \chi = [\chi_1, \chi_2]^T \in \mathbb{R}^2$，$v_d = \sqrt{\dfrac{3}{2}}U_m$，$v_q = 0, U_m v_q = 0$。

则等式（5.6）可以推导为

$$\begin{aligned}\dot{z}_2 &= \bar{a}_1 z_1 + \bar{a}_2 z_2 + bu - \frac{\omega R}{L_q}i_d + \frac{\omega L_d}{L_q}(\varsigma_{12}u + \varsigma_{22}\chi_2) + \varsigma_{11}\dot{u} + \varsigma_{21}\dot{\chi}_1 \\ &= \bar{a}_1 z_1 + \bar{a}_2 z_2 + \underbrace{\left(b + \frac{\omega L_d \varsigma_{12}}{L_q}\right)}_{\tilde{b}}u \underbrace{-\frac{\omega R}{L_q}i_d + \frac{\omega L_d}{L_q}\varsigma_{22}\chi_2 + \varsigma_{11}\dot{u} + \varsigma_{21}\dot{\chi}_1}_{F(x)}\end{aligned} \quad (5.8)$$

其中，x_1 和 x_2 在（5.7）中定义。[①]

[①] Bouc-Wen 模型是描述迟滞非线性的数学唯象模型，其中参数 ς_1 和 ς_2 决定了迟滞非线性的方向。如果 $\varpi_0=2$，$\varpi_1=4$，$\varpi_2=0.5$，$h=3$，$\varsigma_1=3$，$\varsigma_2=5$，输入信号 $u=2.5\sin(1.5\pi)$ 条件下，则 Bouc-Wen 模型的曲线如图 5.1 所示。若参数 $\varsigma_1=-3$ 和 $\varsigma_2=-5$，其他参数不变，则 Bouc-Wen 迟滞曲线如图 5.2 所示。因此 ς_1 和 ς_2 决定了 Bouc-Wen 模型的方向。

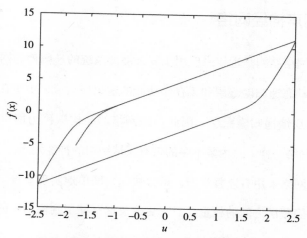

图 5.1　第一种取值下的 Bouc-Wen 模型曲线

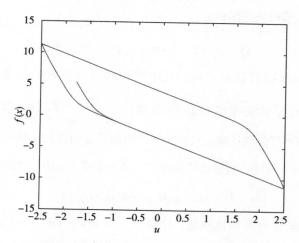

图 5.2　第二种取值下的 Bouc-Wen 模型曲线

根据以上,将系统模型转换为规范形式为

$$\begin{cases} \dot{z}_1 = z_2 \\ \dot{z}_2 = \overline{a}_1 z_1 + \overline{a}_2 z_2 + \overline{b}u + F(x) \\ y = z_1 \end{cases} \quad (5.9)$$

5.3 回声神经网络

回声神经网络结构是近年来应用于复杂动力系统的递归神经网络（RNN）结构。其结构由动态递归隐藏层和无存储功能的输出层组成，由于在结构中增加了像回声一样的反馈递归隐藏层，因此神经网络的结构被称为回声神经网络，与Hopfield 神经网络一样，是为数不多的带有反馈结构的神经网络。

回声神经网络采用有监督学习，可以快速、简单地实现监督学习。另外无需更改输入层和隐藏层之间的权重，只需更改反馈存储层到输出层的权重，回声神经网络即可快速获得高的精度。

首先定义连续时间泄漏积分：

$$\dot{\Omega} = \alpha\left(-\gamma\Omega + \psi\left(W_{in}\nu + W_{it}\Omega + W_{ba}\xi\right)\right) \tag{5.10}$$

其中，Ω 表示 n 维动态状态；$\alpha > 0$ 是时间常数；γ 是神经元集合 Ω 的泄漏衰减率；ψ 表示给定的实连续函数。在本章中，选择为高斯函数。$W_{in} \in \mathbb{R}^{n \times m}$，$W_{it} \in \mathbb{R}^{n \times n}$，$W_{ba} \in \mathbb{R}^{n \times m}$ 分别是输入权重矩阵、内部权重矩阵和反馈权重矩阵。

如文献[29]所述，通过选择 $\alpha=1$ 和 $\gamma=1$，等式（5.10）被改写为

$$\dot{\Omega} = -\Omega + \psi\left(W_{in}\nu + W_{it}\Omega + W_{ba}\xi\right) \tag{5.11}$$

选择高斯函数为激励核函数，Ω 可以定义为 $\Omega(x) = [s_1(x), s_1(x), \ldots, s_l(x)]$，其中 l 表示回声神经网络输出层的神经元的节点。神经网络功能函数选择为

$$s_i(x) = \exp\left\{-\frac{(x-\varrho_i)^T(x-\varrho_i)}{\rho^2}\right\} \tag{5.12}$$

其中，$\varrho_i = [\varrho_{i1}, \varrho_{i1}, \cdots, \varrho_{iq}]^T$，$i = 1, 2, \cdots, q$ 表示输入；ρ 表示高斯函数的宽度，$s_i(x)$ 表示为输出激励函数。

定义输出激励函数 Θ，则回声神经网络的输出为

$$\boldsymbol{\Gamma} = \Theta(\boldsymbol{W}\Omega). \tag{5.13}$$

根据文献[29]，输出激励函数 Θ 单元没有记忆性，因此在时间 $i+1$ 时的值仅部分或间接地取决于其先前值。因此回声神经网络最适合对本质离散时间系统的建模，对连续时间系统反而难以应用，为了解决此问题，可以定义具有连续功能函数，构成连续回声神经网络。

对任意给定的实连续函数 $\boldsymbol{G}(\cdot):\mathbf{R}^n \to \mathbf{R}$，存在足够大的紧集 $\varXi$ 和任意小的 ε_M，使得回声神经网络 $\boldsymbol{\varGamma}$ 满足：

$$\sup_{\Omega \in \varXi} \boldsymbol{G} - \boldsymbol{\varGamma} \leqslant \varepsilon_M, \tag{5.14}$$

其中，$\boldsymbol{G}$ 为给定的实连续函数且存在足够大的紧集 $\varXi$，其上确界为 $\varepsilon_M > 0$。

则连续回声神经网络函数 $\boldsymbol{G}$ 可以表示为

$$\boldsymbol{G} = \boldsymbol{W}^*\Omega(x) + \varepsilon^* \tag{5.15}$$

其中，$\varepsilon^* \leqslant \varepsilon_M$ 是回声神经网络的误差。

5.4 扩张状态观测器

由于永磁同步电机系统考虑了包括磁滞损耗在内的非线性因素，而且部分状态无法直接测得，因此难以直接设计控制器。需要对系统状态进行观测，考虑系统模型规范化形式（5.9），本节将设计一个扩张状态观测器观测系统状态。

在设计扩张状态观测器之前，首先将定义扩张状态 z_3，并将系统扩张为

$$\begin{cases} \dot{z}_1 = z_2 \\ \dot{z}_2 = z_3 + \bar{b}u \\ \dot{z}_3 = \bar{a}_1 z_1 + \bar{a}_2 z_2 + F(x) \end{cases} \quad (5.16)$$

则扩张状态观测器设计为

$$\begin{cases} \dot{\hat{z}}_1 = \hat{z}_2 + \beta_1(y - \hat{z}_1) \\ \dot{\hat{z}}_2 = \hat{z}_3 + \bar{b}u + \beta_2(y - \hat{z}_1) \\ \dot{\hat{z}}_3 = \bar{a}_1 \hat{z}_1 + \bar{a}_2 z_2 + \hat{F}(x) + \beta_3(y - \hat{z}_1) \end{cases} \quad (5.17)$$

其中，$\hat{z}_1$，$\hat{z}_2$，$\hat{z}_3$，$\hat{F}(x)$ 是 z_1，z_2，z_3，$F(x)$ 的估计值，β_1，β_2，β_3 是本节设计的高增益参数。

可将误差定义为

$$\begin{cases} \tilde{z}_i = z_i - \hat{z}, i=1,2,3, \\ \tilde{y} = y - \hat{y}, \\ \tilde{F}(x) = F(x) - \hat{F}(x) \end{cases} \quad (5.18)$$

其中，$F(x)$ 由回声神经网络描述。因此可以得到

$$F(x) = W\Omega(x) + \varepsilon, \quad \hat{F}(x) = \hat{W}\hat{U}(x)$$

所以，有

$$\tilde{F}(x) = \tilde{W}\Omega(x) + \varepsilon = W\Omega(x) + \varepsilon - \hat{W}\Omega(x)$$

成立，且将 $\tilde{W}$ 的更新律定义为

$$\dot{\tilde{W}} = \dot{\hat{W}} = U\left(\hat{e}\Omega(x) - \varepsilon|\hat{e}|\hat{W}\right) \quad (5.19)$$

其中，$U = U^T$ 是一个常对称矩阵，ε 为选取的常正数，e 在（5.36）中定义。根据文献［41］，有如下引理：

引理 5.1 公式（5.19）中的回声神经网络权重 $\tilde{W}$ 由 $\hat{W} \leq \varpi_M/\varepsilon$ 界定，其中 ϖ_M

是回声神经网络基函数向量的边界，即 $\Omega \leqslant \varpi_M$。

证明 由前面的讨论可知，回声神经网络基函数向量 Ω 选取为高斯函数，因此明显有界，即 $\Omega \leqslant \varpi_M$。

选取 Lyapunov 函数为

$$V_N = \frac{1}{2U} \hat{W}^T \hat{W} \tag{5.20}$$

根据式（5.19），可以推导出 V_N 的导数为

$$\dot{V}_N = \frac{1}{U} \hat{W}^T \dot{\hat{W}} = \hat{W}^T \left(\hat{e} \Omega(x) - \varepsilon |\hat{e}| \hat{W} \right) \leqslant -\hat{W} |\hat{e}| \left(\varepsilon \hat{W} - \varpi_M \right) \tag{5.21}$$

根据参考文献[41]，$\hat{W}$ 的边界为 $\hat{W} \leqslant \varpi_M / \varepsilon$，因此 $\tilde{W} = W^* - \hat{W}$ 也被约束为 $\Omega \leqslant \omega_M$，其中，$\omega_M = W_N + (\varpi_M / \varepsilon)$。

根据式（5.16）和式（5.17），有下式成立：

$$\begin{cases} \dot{\tilde{z}}_1 = -\beta_1 \tilde{z}_1 + \tilde{z}_2 \\ \dot{\tilde{z}}_2 = -\beta_2 \tilde{z}_1 + \tilde{z}_3 \\ \dot{\tilde{z}}_3 = (\bar{a}_1 - \beta_3) \tilde{z}_1 + \bar{a}_2 \tilde{z}_2 + \tilde{F}(x). \end{cases} \tag{5.22}$$

则，等式（5.22）可以改写为矩阵形式：

$$\dot{\tilde{z}} = \bar{A} \tilde{z} + \zeta \tilde{F}(x)$$

其中，

$$\bar{A} = \begin{bmatrix} -\beta_1 & 1 & 0 \\ -\beta_2 & 0 & 1 \\ \bar{a}_1 - \beta_3 & \bar{a}_2 & 0 \end{bmatrix}, \tag{5.23}$$

$$\zeta = \begin{bmatrix} 0 \\ 0 \\ 1 \end{bmatrix} \tag{5.24}$$

因为选取合适的高增益参数 β_1，β_2，β_3 可得 A 的特征多项式为 Hurwitz。因此给定一个正定对称矩阵 $P=P^T > 0$，则存在正定对称矩阵 $Q=Q^T > 0$，且满足：

$$\bar{A}^T P + P\bar{A} \leq -Q \text{①} \tag{5.25}$$

定理 5.1 考虑永磁同步电机迟滞模型（5.2）和转换为规范形式的（5.9），其未知迟滞非线性 $F(x)$ 由回声神经网络（5.15）估计，将规范化系统（5.9）扩张为（5.16），假设所有信号有界，则可以通过扩张状态观测器（5.17）观测系统未知状态。

证明 根据扩张状态观测器（5.17）和等式（5.22），将基于扩张状态观测器误差的 Lyapunov 函数 V_0 选取为

$$V_0 = \tilde{z}^T P \tilde{z} \tag{5.26}$$

根据式（5.15），则，V_0 的导数为：

$$\dot{V}_0 = \tilde{z}^T \left(\bar{A}^T P + P\bar{A} \right) \tilde{z} + 2\tilde{F}(x) \zeta^T P \tilde{z} \leq -\tilde{z}^T Q \tilde{z} + 2\tilde{W}\Omega(x) P\tilde{z} + 2\varepsilon \tag{5.27}$$

其中，$\tilde{W} = W - \hat{W}$，$\hat{W}$ 是扩张状态观测器权重的估计值，未知非线性 $\tilde{F}(x)$ 用回声神经网络 $F(x) = W\Omega(x) + \varepsilon$ 估计；因此，$\tilde{F}(x)$ 可得：$\hat{F}(x) = \hat{W}\Omega(x)$。

根据引理 5.1 和参考文献 [31]，扩张状态观测器的误差权重满足 $\tilde{W} \leq w_M$，其中 $\Omega(x)$ 有界，且有 $\Omega(x) \leq \varpi_M$。考虑到 $\varepsilon \leq \varepsilon_M$，则公式（5.27）表示为：

---

① 大多数情况下扩张状态观测器被用来观测系统总扰动，当作扰动观测器使用。但是在本章中，调整了扩张状态观测器的结构，作为状态观测器使用，观测系统状态。若选择合适的高增益 β_1，β_2，β_3，则扩张状态观测器可以快速而精确地估计系统状态。另外，可容易调整扩张状态观测器保证状态观测的收敛性。同时为了简化扩张状态观测器设计，本节使用回声神经网络估计未知迟滞非线性和扰动，具体证明过程见定理 5.1。

$$\dot{V}_0 \leqslant -\tilde{z}^T Q \tilde{z} + 2w_M \varphi_M \lambda_{\max}(P)\tilde{z} + 2\varepsilon_M \leqslant -\tilde{z}^T Q \tilde{z} + 2w_M \varphi_M \lambda_{\max}(P)\tilde{z} + 2\varepsilon_M \quad (5.28)$$

其中，$\lambda_{\max}(P)$ 表示矩阵 P 的最大范数。根据杨氏不等式 $ab \leqslant (a^2+b^2)/2$，可以得到：$2w_M \varphi_M \lambda_{\max}(P)\tilde{z} \leqslant w_M^2 + \varphi^2 + \lambda_{\max}^2(P) + \tilde{z}^2$。因此有下式成立：

$$\dot{V}_0 \leqslant -\tilde{z}^T Q \tilde{z} + w_M^2 + \varphi^2 + \lambda_{\max}^2(P) + \tilde{z}^2 + 2\varepsilon_M \quad (5.29)$$

将（5.29）写为

$$\dot{V}_0 \leqslant -a_1 V_0 + a_2 \quad (5.30)$$

其中，a_1 可以通过 $\bar{A}^T P + P\bar{A} \leqslant -Q$ 推导获得，$a_2 = w_M^2 + \varphi^2 + \lambda_{\max}^2(P) + \tilde{z}^2 + 2\varepsilon_M$。对（5.30）的两边积分，则以下不等式成立：

$$|\tilde{z}| \leqslant \sqrt{\frac{V_0(0)e^{-a_1 t} + (a_2/a_1)}{\lambda_{\max}(P)}} \quad (5.31)$$

因此可知扩张状态观测器中所有信号有界，状态观测误差 $\tilde{z}$ 收敛于平衡点附近的小邻域内。因此，根据李雅普诺夫理论，所提出的扩张状态观测器可观测系统状态。

5.5 滑模控制器设计

本节将设计永磁同步电机迟滞系统的自适应滑模控制器。在 5.4 节中设计了扩张状态观测器用来观测难以直接测得的系统状态，对于未知迟滞部分和其他干扰项采用回声神经网络估计补偿。本节在此基础上设计自适应滑模控制来控制该永磁同步电机系统。控制结构如图 5.3 所示。首先定义一个滑模面，然后根据提出的设计滑模控制器，最后证明控制策略的稳定性。

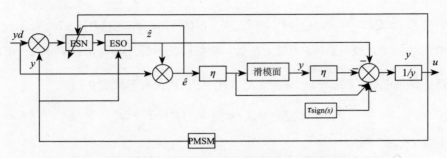

图 5.3 带回声神经网络和扩张状态观测器的滑模控制器结构

为了设计滑模控制，定义跟踪误差为

$$e = y - y_d = z_1 - y_d \tag{5.32}$$

设计滑模面为

$$s = \eta e + \dot{e} \tag{5.33}$$

其中，设计参数 $\eta > 0$ 且参考输入信号 y_d 具有连续导数。

根据（5.32），s 的导数为

$$\dot{s} = \eta \dot{e} + \ddot{e} \tag{5.34}$$

考虑式（5.32）定义的跟踪误差，可得：

$$\dot{e} = \dot{z}_1 - \dot{y}_d = z_2 - \dot{y}_d, \quad \ddot{e} = \dot{z}_2 - \ddot{y}_d \tag{5.35}$$

定义参数误差和自适应率为

$$\hat{e} = \hat{y} - y_d = \hat{z}_1 - y_d, \quad \hat{s} = \eta \hat{e} + \dot{\hat{e}}, \quad \tilde{s} = s - \hat{s}, \quad \tilde{e} = e - \hat{e} \tag{5.36}$$

则，滑模控制器设计为

$$u = \frac{1}{b}\left(-\eta_t \hat{s} - \hat{z}_3 - \eta \dot{\hat{e}} + \ddot{y}_d - \tau \operatorname{sign}(s)\right) \tag{5.37}$$

其中，设计参数 $\eta_t > 0$，$\hat{s}$ 和 $\hat{e}$ 分别表示 s 和 e 的估计。①

得到如下定理：

定理 5.2 对于考虑磁滞损耗的永磁同步电机模型（5.9），将其状态扩张为（5.16），对于不能直接测得的系统状态采用扩张状态观测器（5.17）观测，定义滑模面（5.33），设计滑模控制器（5.37），则闭环系统中的所有信号都将最终一致有界（UUB）。

证明 选择 Lyapunov 函数为

$$V = \frac{1}{2}s^2 \tag{5.38}$$

根据前面定义（5.33）～（5.35），则 V 的导数为：

$$\dot{V} = s\dot{s} = s\left(\eta \dot{e} + \dot{e}\right) = s\left(\eta \dot{e} + \left(\dot{z}_2 - \ddot{y}_d\right)\right) \tag{5.39}$$

将（5.16）中的 $\dot{z}_2 = z_3 + \bar{b}u$ 代入到（5.39）中，得到：

$$\dot{V} = s\left(\eta \dot{e} + \left(z_3 + \bar{b}u - \ddot{y}_d\right)\right) \tag{5.40}$$

将滑模控制器（5.37）代入（5.40），可得 V 的导数为

$$\dot{V} = s\left(\eta \dot{e} + \left(z_3 - \eta_t \hat{s} - \hat{z}_3 - \eta \hat{e} + \ddot{y}_d - \tau\, \mathrm{sign}(s) - \ddot{y}_d\right)\right) \tag{5.41}$$

将（5.36）代入（5.41），可得：

---

① 为了降低"抖震"的影响，选择 $0 < \tau < 1$。但是由于符号函数是不连续函数，"抖震"仍然影响控制精度，本章在证明时仍然采用符号函数，但是在仿真时为了进一步降低"抖震"的影响，将符号函数 sign(·) 用双曲正切函数 tanh(·) 代替。

$$\dot{V} = s\left(\eta\tilde{e} + \tilde{z}_3 - \eta_t\hat{s} - \tau\mathrm{sign}(s)\right) = s\left(\eta\tilde{e} + \tilde{z}_3 - \eta_t(s - \tilde{s}) - \tau\mathrm{sign}(s)\right)$$
$$= -\eta_t s^2 + s\left(\eta\tilde{e} + \tilde{z}_3 + \eta_t\tilde{s} - \tau\mathrm{sign}(s)\right) \tag{5.42}$$

将滑模面（5.33）和（5.36）中的 $\hat{s} = \eta\hat{e} + \dot{\hat{e}}$ 代入 $\tilde{s}$ 得到：

$$\tilde{s} = \eta\tilde{e} + \dot{\tilde{e}} \tag{5.43}$$

然后将（5.43）代入（5.42），可得：

$$\dot{V} = -\eta_t s^2 + s\left(\eta\tilde{e} + \tilde{z}_3 + \eta\eta_t\tilde{e} + \eta_t\dot{\tilde{e}} - \tau\mathrm{sign}(s)\right) = -\eta_t s^2 + s\left((\eta + \eta_t)\dot{\tilde{e}} + \tilde{z}_3 + \eta\eta_t\tilde{e} - \tau\mathrm{sign}(s)\right) \tag{5.44}$$

根据（5.32）中的 $e = z_1 - y_d$，有下式成立：

$$\tilde{e} = \tilde{z}_1 - y_d$$
$$\dot{\tilde{e}} = \tilde{z}_2 - \dot{y}_d \tag{5.45}$$

将（5.22）和（5.45）代入到（5.44）中，得：

$$\dot{V} = -\eta_t s^2 + s\left((\eta + \eta_t)\tilde{z}_2 - (\eta + \eta_t)\dot{y}_d + \bar{a}_1\tilde{z}_1 + \bar{a}_2\tilde{z}_2 + \tilde{F}(x) + \eta\eta_t\tilde{z}_1 - \eta\eta_t y_d - \tau\mathrm{sign}(s)\right)$$
$$= -\eta_t s^2 + s\left((\bar{a}_1 + \eta\eta_t)\tilde{z}_1 + (\bar{a}_2 + \eta + \eta_t)\tilde{z}_2 - \eta\eta_t y_d - (\eta + \eta_t)\dot{y}_d + \tilde{F}(x) - \tau\mathrm{sign}(s)\right) \tag{5.46}$$

又根据定理 1，未知的有界函数 $\tilde{F}(x)$ 可以用 $\tilde{F}(x) = \tilde{W}\Omega(x) \leq \omega_M\varphi_M$ 来近似，且 $\tilde{z}_1$, $\tilde{z}_2$, y_d 都是有界的。定义 $(\bar{a}_1 + \eta\eta_t)\tilde{z}_1 + (\bar{a}_2 + \eta + \eta_t)\tilde{z}_2 - \eta\eta_t y_d - (\eta + \eta_t)\dot{y}_d + \tilde{W}(x) \leq \Delta_{max}$，其中 $\Delta_{max} \geq 0$，则等式（5.46）满足：

$$\dot{V} \leq -\eta_t s^2 + s\Delta_{max} - \tau s\,\mathrm{sign}(s) \leq -\eta_t s^2 + s\Delta_{max} - \tau|s| \tag{5.47}$$

若 $\tau|s| > 1$，方程（5.47）可改写为

$$\dot{V} \leqslant -\eta_t s^2 + s\Delta_{\max} = -\mu + s\Delta_{\max} \quad (5.48)$$

其中，$\mu = \eta_t > 0$

若 $0 \leqslant \tau|s| \leqslant 1$，则

$$\dot{V} \leqslant -\eta_t s^2 + s\Delta_{\max} - \tau s^2 = (\eta_t + \tau)s^2 + s\Delta_{\max} = -\mu s^2 + s\Delta_{\max} \quad (5.49)$$

其中，$\mu = \eta_t + \tau > 0$。

根据杨氏不等式，$s\Delta_{\max}$ 满足 $s\Delta_{\max} \leqslant (s^2/2) + (\Delta_{\max}^2/2)$，将等式（5.47）写为：

$$\dot{V} \leqslant -\mu s^2 + \frac{s^2}{2} + \frac{\Delta_{\max}^2}{2} = \left(-\mu + \frac{1}{2}\right)s^2 + \frac{\Delta_{\max}^2}{2} \quad (5.50)$$

当 $\mu > 0.5$ 时，根据（5.38）中的 Lyapunov 函数 V 的定义，解方程（5.50）得到：

$$|s| \leqslant \sqrt{\frac{(2\mu-1)V(0) + \Delta_{\max}^2}{(2\mu-1)\Delta_{\max}}} \quad (5.51)$$

考虑（5.33）中 s 的定义和（5.51）中有界的定义，可求得到误差 e 为

$$|e| \leqslant \sqrt{\frac{s(0) + (\eta/\vartheta)}{\vartheta}} \quad (5.52)$$

其中，$\vartheta = \sqrt{((2\mu-1)V(0) + \Delta_{\max}^2)/((2\mu-1)\Delta_{\max})}$

根据李雅普诺夫理论，闭环的所有信号最终一致有界，且 s,e 的边界如（5.51）（5.52）。

得证。

5.6 仿真

本节使用 Matlab 的 Simulink 模块进行仿真，验证所提出的扩张状态观测器和自适应滑模控制器对考虑磁滞损失的永磁同步电机系统控制的有效性。永磁同步电机、扩张状态观测器和自适应滑模控制器的参数选择如下：

d 轴电感 L_d 和 q 轴电感 L_q 为 0.47Mh；扭矩 T 是 96.5N·m；角速度 ω 为 3000rad/s；电阻 R 为 0.033Ω 电机感应电压常数 K_l 为 0.147；电机转矩常数 K_t 为 0.72。高增益参数 β_1、β_2、β_3 选为 $\beta_1=360, \beta_2=5000, \beta_3=28500$。$\eta$ 选为 5，τ 可选为 0.3。为了验证所提出的方法，在仿真中将误差图的边界规定为 [-0.5,0.5]。

在本节仿真中，为了对比采用双曲正切函数代替符号函数对滑模控制器的"抖震"抑制作用，设计两个不同的 $\tau\mathrm{sign}(s)$ 进行仿真。其中令一个 $\tau\mathrm{sign}(s)=0.3\mathrm{sign}(s)$，令另一个 $\tau\mathrm{sign}(s)=0.3\mathrm{tanh}(s)$。在这两个选择下，其他参数保持不变，对比两个选择对滑模控制器"抖震"的不同抑制作用。

5.6.1 仿真1

考虑自适应滑模控制器（5.37），选择 $\tau\mathrm{sign}(s)=0.3\mathrm{sign}(s)$，验证本章所提出的回声神经网络补偿迟滞非线性和其他扰动，扩张状态观测器观测系统状态以及滑模控制器控制永磁同步电机迟滞系统的效果。跟踪输出的结果 y, $\dot{y}$, 和 $\ddot{y}$ 如图 5.4 所示，跟踪误差如图 5.5 所示。扩张状态观测器的观测结果 z_1, z_2, z_3 如图 5.6 所示，观测误差如图 5.7 所示。图 5.8 是滑模控制器的控制输入。

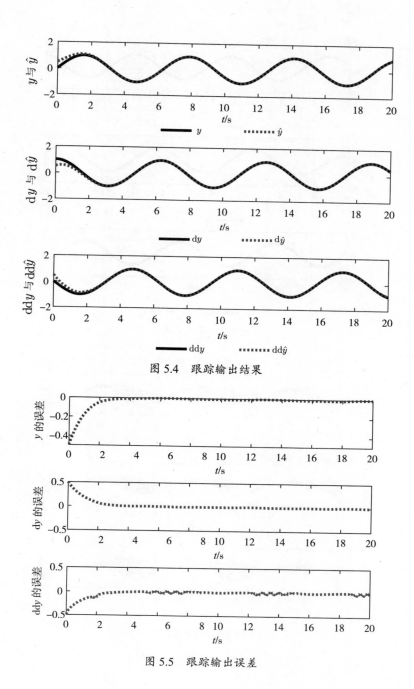

图 5.4 跟踪输出结果

图 5.5 跟踪输出误差

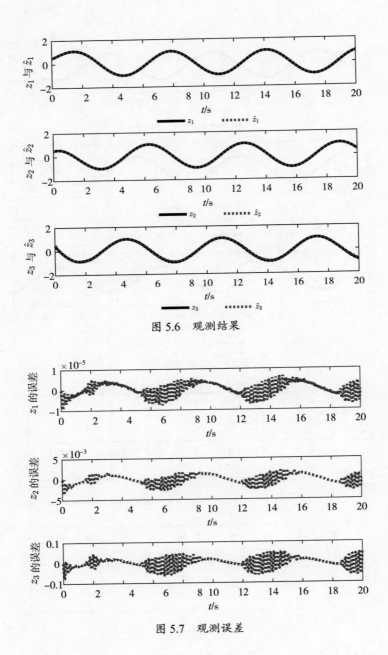

图 5.6 观测结果

图 5.7 观测误差

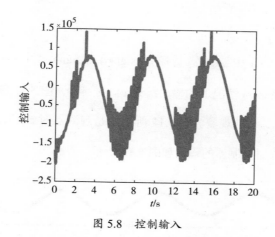

图 5.8 控制输入

从图 5.4 和图 5.6 可以明显看出，本章所提出的扩张状态观测器和自适应滑模控制器具有很好的效果，控制器很好地跟踪了参考信号，观测器很好地观测了系统状态。从图 5.3 控制器的结构可知，扩张状态观测器以及回声神经网络的调整是在闭环内在线进行的，并且随控制器的调整而调整。如果扩张状态观测器的 β_1、β_2 和 β_3 的增益选择得足够高，它可以快速逼近状态 z_1、z_2 和 z_3。从图 5.6 中，可以清楚地看到扩张状态观测器运行良好，但是从图 5.5 和图 5.7 中可知，由于滑模控制器的"抖震"问题，永磁同步电机虽然在输出时是平滑稳定的，但是其状态在换向时抖动明显。观察图 5.8 也可以看出滑模控制器的输入在换向时进行了剧烈调节，与观测器的误差具有类似的特征。

从图 5.7～5.8 中看出即使回声神经网络补偿了迟滞非线性和其他扰动，滑模控制器仍然受到了"抖震"的影响，滑模控制的"抖震"确实降低了控制策略的精度。为了降低滑模控制器"抖震"的影响，在下面选择了连续的函数 tanh(.) 来代替不连续的 sign(.) 函数进行验证。

5.6.2 仿真2

将控制器（5.37）中 $\tau\mathrm{sign}(s)$ 选择为 $\tau\mathrm{sign}(s)=0.3\tanh(s)$。图 5.9 和图 5.10 显示为在 $\tau\mathrm{sign}(s)=0.3\tanh(s)$ 情况下的输出结果 y，$\dot{y}$，和 $\ddot{y}$，图 5.11 显示了扩张状态观测器对状态 z_1，z_2，z_3 的观测。图 5.12 所示的是观测器对 z_1，z_2，z_3 的估计误差。$\tau\mathrm{sign}(s)=0.3\tanh(s)$ 时控制器的输入如图 5.13 所示。

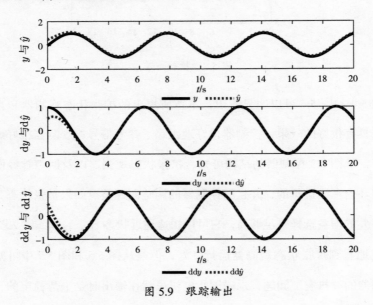

图 5.9 跟踪输出

第 5 章 Bouc-Wen 迟滞非线性系统扩张状态观测器与回声神经网络滑模控制

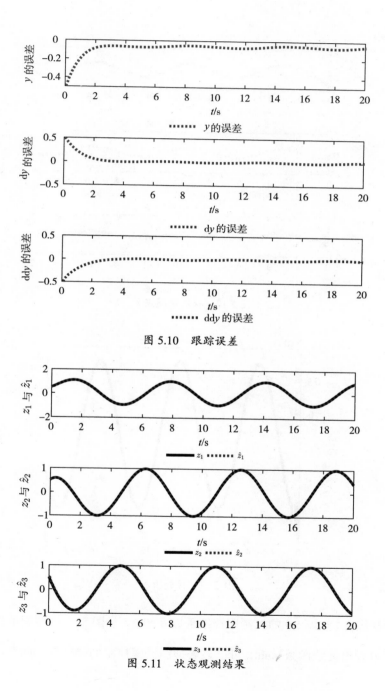

图 5.10 跟踪误差

图 5.11 状态观测结果

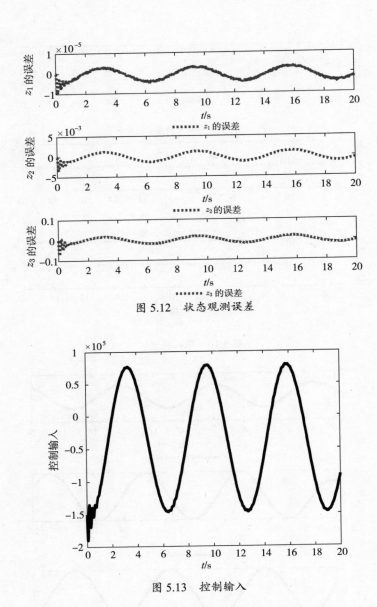

图 5.12 状态观测误差

图 5.13 控制输入

根据图 5.4 和图 5.9 的控制跟踪结果以及图 5.5 和图 5.10 的控制器误差相比较，可以看出连续函数 tanh(.) 可以显著抑制滑模控制的"抖震"问题。特别是与图 5.5 和图 5.10 的跟踪误差相比，采用连续的双曲正切函数 $\tau\mathrm{sign}(s)=0.3\tanh(s)$ 作为鲁棒项时，"抖震"的影响微乎其微。从图 5.11 和图 5.12 中的观测结果和误差也可以得出相同结论。特别是图 5.7 和图 5.12 对比可知，在图 5.7 的换向过程中 z_3 最大误差在 0.08 左右，而图 5.12 中 z_3 的最大误差仅仅约为 0.02。另外，图 5.12 中用连续的双曲正切函数 $\tau\mathrm{sign}(s)=0.3\tanh(s)$ 作为鲁棒项时，z_1，z_2，z_3 的最大误差出现在暂态过程中，这是由回声神经网络引起的，但在图 5.8 用不连续符号函数 $\tau\mathrm{sign}(s)=0.3\mathrm{sign}(s)$ 时，z_1，z_2，z_3 的最大误差出现在换向过程中，表明是由滑模控制器的"抖震"引起的。

表 5.1　$\tau\mathrm{sign}(s)=0.3\mathrm{sign}(s)$ 时跟踪误差和观测误差

	绝对平均误差	最大误差	最小误差
$\tilde{y}$	0.0276	0.5	0.0014
$\dot{\tilde{y}}$	0.0269	0.5	$1.8734\,e^{-6}$
$\ddot{\tilde{y}}$	0.0317	0.5	$1.3852\,e^{-6}$
$\tilde{z}_1$	$2.702e^{-6}$	$9.9556e^{-6}$	0
$\tilde{z}_2$	$9.5797e^{-6}$	0.0035	0
$\tilde{z}_3$	0.0187	0.0763	0

表 5.2　$\tau\mathrm{sign}(s)=0.3\tanh(s)$ 误差和观测误差

	绝对平均误差	最大误差	最小误差
$\tilde{y}$	0.0775	0.5	0.047
$\dot{\tilde{y}}$	0.0281	0.5	$7.9464e^{-6}$
$\ddot{\tilde{y}}$	0.0312	0.5	$1.473e^{-6}$
$\tilde{z}_1$	$2.2891e^{-6}$	$8.3276e^{-6}$	0
$\tilde{z}_2$	$8.3078e^{-6}$	0.0032	0
$\tilde{z}_3$	0.0116	0.0698	0

表 5.1 是采用符号函数 $\tau\mathrm{sign}(s)=0.3\mathrm{sign}(s)$ 作为控制器鲁棒项时的跟踪误差和观测器误差，采用连续双曲正切函数 $\tau\mathrm{sign}(s)=0.3\tanh(s)$ 时的跟踪误差和观测器误差如表 5.2 所示。

从表 5.1 和表 5.2 可以看出，$[\tilde{z}_1, \tilde{z}_2, \tilde{z}_3]^T$ 的平均绝对误差分别为 $[2.702e^{-6}, 9.5797e^{-4}, 0.0187]^T$ 和 $[2.289e^{-6}, 8.3078e^{-4}, 0.0116]^T$。结果表明，连续双曲正切函数 tanh(.) 可以显著抑制滑模控制器的"抖震"。但是考虑到滑模控制器的平均绝对误差 $\left[\tilde{y}, \tilde{\dot{y}}, \tilde{\ddot{y}}\right]^T$ 分别为 $[0.0276, 0.0269, 0.0317]^T$ 和 $[0.0775, 0.0281, 0.0312]^T$，最小跟踪误差分别为 $[0.0014, 1.8734e^{-6}, 1.3852e^{-5}]$ 和 $[0.047, 7.9464e^{-6}, 1.473e^{-5}]$。这表明连续双曲正切函数 tanh(.) 虽然可以显著抑制滑模控制器的"抖震问题"，但是"抖震"是滑模控制器得以运行的内部驱动力，过于抑制不但降低到达滑模面的时间，还会降低跟踪精度。

图 5.8 和 5.13 相比，可以明显地发现在图 5.13 中暂态过程受到回声神经网络的影响，稳态过程的调节则比较平滑，说明连续的双曲正切函数 tanh(.) 作为鲁棒项抑制了滑模控制器的"抖震"问题。但是采用不连续的符号函数 sign(.) 作为鲁棒项时难以达到图 5.13 那样的稳态过程。

5.7 总结

本章针对汽车用永磁同步电机系统系统进行研究，首先考虑了磁滞等影响建立了新的永磁同步电机迟滞模型。为了更好地简化状态观测器和控制器设计，将提出的新的模型转换为规范形式。考虑到系统部分状态无法直接测得，设计了扩张状态观测器来观测未知状态，同时采用回声神经网络估计补偿未知迟滞非线性和其他扰动以简化观测器的设计。最后提出了自适应滑模控制器控制考虑迟滞的永磁同步电机系统，并应用连续双曲正切函数代替不连续的符号函数作为控制器的鲁棒项，降低滑模控制器中"抖震"的影响。

参考文献

[1] NA J, CHEN AS, HERRMANN G, et al. Vehicle engine torque estimation via unknown input observer and adaptive parameter estimation [J]. IEEE Transactions on Vehicular Technology, 2018, 67(1):409–422.

[2] NA J, HUANG YB, WU X, et al. Active adaptive estimation and control for vehicle suspensions with prescribed performance [J]. IEEE Transactions on Control Systems Technology, 2018, 26(6): 2063–2077.

[3] WANG SB, YU HS, YU JP, et al. Neural-networkbasedadaptive funnel control for servo mechanisms with unknown dead-zone [J]. IEEE Transactions on Cybernetics, 2018: 1–12.

[4] WANG SB, NA J, REN XM, et al. Unknown input observer-based robust adaptive funnel motion control for nonlinear servome-chanisms [J]. International Journal of Robust and Nonlinear Control, 2018,28(18): 6163–6179.

[5] HUANG YB, NA J, WU X, et al. Approximation-free control for vehicle active suspensions with hydraulic actuator [J]. IEEE Transactions on Industrial Electronics, 2018,65(9): 7258–7267.

[6] NA J, HUANG YB, WU X, et al. Adaptive finitetimefuzzy control of nonlinear active suspension systems with input delay [J]. IEEE Transactions on Cybernetics, 2019: 1–12.

[7] HU C, GAO H, GUO J. MME-EKF-based path-tracking control of autonomous vehicles considering input saturation [J]. IEEE Transactions on Vehicular Technology, 2019,68(6): 5246–5259.

[8] LI B, DU H, LI W, et al. Integrated trajectory planning and control for obstacle avoidance manoeuvre using nonlinear vehicle mp algorithm," [J]. IET Intelligent Transport Systems, 2019, 13(2): 385–397.

[9] SUN S, DENG H, DU H. A compact variable stiffness and damping shock absorber for vehicle suspension [J]. IEEE/ASME Transactions on Mechatronics, 2015,20(5): 2621–2629.

[10] SUN S, TANG X, YANG J. A new generation of magnetor-heological vehicle suspension system with tunable stiffness and damping characteristics [J]. IEEE Transactions on Industrial Informatics, 2019, 15(8): 4696–4708.

[11] LUO G, ZHANG R, CHEN Z, et al. Anovel nonlinear modeling method for permanent-magnet synchronous motors [J]. IEEE Transactions on Industrial Electronics, 2016, 63(10): 6490–6498.

[12] EGOROV D, PETROV I, LINK J, et al. Model-based hysteresis loss assessment in PMSMs with ferrite magnets [J]. IEEE Transactions on Industrial Electronics, 2018,65(1): 179–188.

[13] LIN X, HUANG W, WANG L. SVPWM strategy based on the hysteresis controller of zero-sequence current for threephase open-end winding PMSM[J]. IEEE Transactions on Power Electronics, 2019, 34(4): 3474–3486.

[14] GAO XH. Adaptive neural control for hysteresis motor drivingservo system with Bouc-Wen model [J]. Complexity, 2018: 1-9.

[15] GAO XH, LIU RG. Multiscale Chebyshev neural network Identification and adaptive control for backlash-like hysteresis system [J]. Complexity, 2018: 1-9.

[16] ZHU Q, ZHANG W, NA J, et al. U-model based control Design framework for continuous-time systems [C]. Proceedings of the 2019 Chinese Control Conference (CCC), Guangzhou, China, 2019: 106–111.

[17] ZHANG J, ZHU Q, LI Y, et al. A general U-model based super twisting design procedure for nonlinear polynomial systems [C]. Proceedings of the 2017 36th Chinese Control Conference (CCC), Dalian, China, 2017: 3641–3645.

[18] ZHU Q, GUO L. Stable adaptive neurocontrol for nonlineardiscrete-time systems [J]. IEEE Transactions on Neural Networks, 2004,15(3): 653–662.

[19] ZHU Q, ZHAO D, ZHANG S, et al. U-model enhanceddynamic control of a heavy oil pyrolysis/cracking furnace [J]. IEEE/CAA Journal of Automatica Sinica, 2018, 5(2): 577–586.

[20] WEI W, WANG M, LI D, et al. Disturbanceobserver based active and adaptive synchronization of energy resource chaotic system [J]. ISA Transactions, 2016(65): 164–173.

[21] HE W, MENG T, HE X, et al. Iterative learning controlfor a flapping wing micro aerial vehicle under distributed disturbances [J]. IEEE Transactions on Cybernetics, 2019, 49(4): 1524–1535.

[22] ZHU Y, ZHENG WX. Observer-based control for cyberphysicalsystems with periodic dos attacks via a cyclic switching strategy [J]. IEEE Transactions on Automatic Control, 2019,99: 1.

[23] ZHU Y, ZHENG WX, ZHOU D. Quasi-synchronization of discrete-time Lur'e type switched systems with parameter mismatches and relaxed PDT constraints [J]. IEEE Transactions on Automatic Control, 2019:1–12.

[24] ZHU Y, ZHENG WX. Multiple Lyapunov functions analysis approach for discrete-time switched piecewise affine systems under dwell-time constraints[J]. IEEE Transactions on Automatic Control, 2019(99): 1.

[25] HAN JQ. The extended state observer for a class of uncertainsystems [J]. Control and Decision, 1995, 10(1):85–88.

[26] XUE WC, BAI W, YANG S, et al. ADRC with adaptive extended state observer and its application to airfuel ratio control in gasoline engines [J]. IEEE Transactions on Industrial Electronics, 2015, 62(9): 5847–5857.

[27] XUE WC, MADONSKI R, LAKOMY K, et al. Add-on module of active disturbance rejection for set-point tracking of motion control systems [J]. IEEE Transactions on Industry Applications, 2017, 53(4): 4028–4040.

[28] CHEN Q, TAO L, NAN Y. Full-order sliding mode controlfor high-order nonlinear system based on extended state observer [J]. Journal of Systems Science and Complexity, 2016, 29(4): 978–990.

[29] SUN GF, REN XM, LI D. Neural active disturbance rejectionoutput control of

multimotor servomechanism [J]. IEEE Transactions on Control Systems Technology, 2015, 23(2): 746–753.

[30] WANG SB, REN XM, NA J, et al. Extended-state-observerbasedfunnel control for nonlinear servomechanisms with prescribed tracking performance [J]. IEEE Transactions on Automation Science and Engineering, 2017, 14(1): 98–108.

[31] CHEN Q, SHI L, NA J, et al. Adaptive echo state network control for a class of pure-feedback systems with input and output constraints [J]. Neurocomputing, 2018(275): 1370–1382.

[32] SUN GF, LI D, REN XM. Modified neural dynamic surfaceapproach to output feedback of MIMO nonlinear systems [J]. IEEE Transactions on Neural Networks and Learning Systems, 2015, 26(2): 224–236.

[33] WANG SB, NA J, REN XM. Rise-based asymptotic prescribedperformance tracking control of nonlinear servo mechanisms [J]. IEEE Transactions on Systems, Man, and Cybernetics: Systems, 2018, 48(12): 2359–2370.

[34] WANG ML, REN XM, CHEN Q, et al. Modifieddynamic surface approach with bias torque for multi-motor servo-mechanism [J]. Control Engineering Practice, 2016(50): 57–68.

[35] WANG SB, REN XM, NA J, et al. Robust tracking and vibration suppression for nonlinear two-inertia system viamodified dynamic surface control with error constraint [J]. Neurocomputing, 2016(203): 73–85.

[36] GAO XH, ZHANG CY, ZHU CS, et al. Identification and control for hammerstein systems with hysteresis non-linearity [J]. IET Control eory & Applications, 2015, 9(13): 1935–1947.

[37] TAO L, CHEN Q, NAN Y, et al. Double hyperbolic reaching law with chattering-free and fast convergence [J]. IEEE Access, 2018(6): 27717–27725.

[38] ZHAO W, REN XM, WANG SB. Parameter estimation-based time-varying sliding mode control for multimotor driving servo systems [J]. IEEE/ASME Transactions on Mechatronics, 2017, 22(5): 2330–2341.

[39] Chen Q, Xie S, Sun MX, et al. Adaptive nonsingular fixed-time attitude stabilization of uncertain spacecraft [J]. IEEE Transactions on Aerospace and Electronic Systems, 2018, 54(6): 2937–2950.

[40] BOGOSYAN S, GOKASAN M, GOERING DJ. A novel model validation and estimation approach for hybrid serial electric vehicles [J]. IEEE Transactions on Vehicular Technology, 2007, 56(4): 1485–1497.

[41] NA J, REN XM, ZHENG DD. Adaptive control for nonlinear pure-feedback systems with high-order sliding mode observer [J]. IEEE Transactions on Neural Networks and Learning Systems, 2013, 24(3): 370–382.

致　谢

　　感谢我的导师北京理工大学任雪梅教授的悉心指导！感谢学长学弟们及其他帮助过我的人！感谢吕昊、万裕、王剑昊、刘震等同学在排版及公式录入过程中给与的帮助！感谢编辑们的辛苦工作与帮助！